Advanced Ceramic Coatings and Interfaces IV

Advanced Ceramic Coatings and Interfaces IV

A Collection of Papers Presented at the 33rd International Conference on Advanced Ceramics and Composites
January 18–23, 2009
Daytona Beach, Florida

Edited by
Dongming Zhu
Hua-Tay Lin

Volume Editors
Dileep Singh
Jonathan Salem

A John Wiley & Sons, Inc., Publication

Published by John Wiley & Sons, Inc., Hoboken, New Jersey.
Published simultaneously in Canada.

Library of Congress Cataloging-in-Publication Data is available.

ISBN 978-0-470-45753-5

Contents

Preface

The Symposium on Advanced Ceramic Coatings for Structural, Environmental and Functional Applications was held at the 33rd Cocoa Beach International Conference on Advanced Ceramics and Composites in Cocoa Beach, Florida, during January 18-23, 2009. A total of 70 papers, including 8 invited talks, were presented at the symposium, covering broad ceramic coating and interface topic areas and emphasizing the latest advancement in coating processing, characterization and development.

The present volume contains twelve contributed papers from the symposium, with topics including vibration damping coatings, thermal and environmental barrier coating processing, testing and life modeling, non-destructive evaluation, multifunctional coatings and interfaces, highlighting the state-of-the-art ceramic coatings technologies for various critical engineering applications.

We are greatly indebted to the members of the symposium organizing committee, including Uwe Schulz, Yutaka Kagawa, Rodney Trice, Irene T. Spitsberg, Dileep Singh, Robert Vassen, Sophoclis Patsias, Yong-Ho Sohn, Anette M. Karlsson, and Ping Xiao, for their assistance in developing and organizing this vibrant and cutting-edge symposium. We also would like to express our sincere thanks to manuscript authors and reviewers, all the symposium participants and session chairs for their contributions to a successful meeting. Finally, we are also grateful to the staff of The American Ceramic Society for their efforts in ensuring an enjoyable conference and the high-quality publication of the proceeding volume.

DONGMING ZHU
H. T. LIN

Introduction

The theme of international participation continued at the 33rd International Conference on Advanced Ceramics and Composites (ICACC), with over 1000 attendees from 39 countries. China has become a more significant participant in the program with 15 contributed papers and the presentation of the 2009 Engineering Ceramic Division's Bridge Building Award lecture. The 2009 meeting was organized in conjunction with the Electronics Division and the Nuclear and Environmental Technology Division.

Energy related themes were a mainstay, with symposia on nuclear energy, solid oxide fuel cells, materials for thermal-to-electric energy conversion, and thermal barrier coatings participating along with the traditional themes of armor, mechanical properties, and porous ceramics. Newer themes included nano-structured materials, advanced manufacturing, and bioceramics. Once again the conference included topics ranging from ceramic nanomaterials to structural reliability of ceramic components, demonstrating the linkage between materials science developments at the atomic level and macro-level structural applications. Symposium on Nanostructured Materials and Nanocomposites was held in honor of Prof. Koichi Niihara and recognized the significant contributions made by him. The conference was organized into the following symposia and focused sessions:

Symposium 1	Mechanical Behavior and Performance of Ceramics and Composites
Symposium 2	Advanced Ceramic Coatings for Structural, Environmental, and Functional Applications
Symposium 3	6th International Symposium on Solid Oxide Fuel Cells (SOFC): Materials, Science, and Technology
Symposium 4	Armor Ceramics
Symposium 5	Next Generation Bioceramics
Symposium 6	Key Materials and Technologies for Efficient Direct Thermal-to-Electrical Conversion
Symposium 7	3rd International Symposium on Nanostructured Materials and Nanocomposites: In Honor of Professor Koichi Niihara
Symposium 8	3rd International symposium on Advanced Processing & Manufacturing Technologies (APMT) for Structural & Multifunctional Materials and Systems

Symposium 9	Porous Ceramics: Novel Developments and Applications
Symposium 10	International Symposium on Silicon Carbide and Carbon-Based Materials for Fusion and Advanced Nuclear Energy Applications
Symposium 11	Symposium on Advanced Dielectrics, Piezoelectric, Ferroelectric, and Multiferroic Materials
Focused Session 1	Geopolymers and other Inorganic Polymers
Focused Session 2	Materials for Solid State Lighting
Focused Session 3	Advanced Sensor Technology for High-Temperature Applications
Focused Session 4	Processing and Properties of Nuclear Fuels and Wastes

The conference proceedings compiles peer reviewed papers from the above symposia and focused sessions into 9 issues of the 2009 Ceramic Engineering & Science Proceedings (CESP); Volume 30, Issues 2-10, 2009 as outlined below:

- Mechanical Properties and Performance of Engineering Ceramics and Composites IV, CESP Volume 30, Issue 2 (includes papers from Symp. 1 and FS 1)
- Advanced Ceramic Coatings and Interfaces IV Volume 30, Issue 3 (includes papers from Symp. 2)
- Advances in Solid Oxide Fuel Cells V, CESP Volume 30, Issue 4 (includes papers from Symp. 3)
- Advances in Ceramic Armor V, CESP Volume 30, Issue 5 (includes papers from Symp. 4)
- Advances in Bioceramics and Porous Ceramics II, CESP Volume 30, Issue 6 (includes papers from Symp. 5 and Symp. 9)
- Nanostructured Materials and Nanotechnology III, CESP Volume 30, Issue 7 (includes papers from Symp. 7)
- Advanced Processing and Manufacturing Technologies for Structural and Multifunctional Materials III, CESP Volume 30, Issue 8 (includes papers from Symp. 8)
- Advances in Electronic Ceramics II, CESP Volume 30, Issue 9 (includes papers from Symp. 11, Symp. 6, FS 2 and FS 3)
- Ceramics in Nuclear Applications, CESP Volume 30, Issue 10 (includes papers from Symp. 10 and FS 4)

The organization of the Daytona Beach meeting and the publication of these proceedings were possible thanks to the professional staff of The American Ceramic Society (ACerS) and the tireless dedication of the many members of the ACerS Engineering Ceramics, Nuclear & Environmental Technology and Electronics Divisions. We would especially like to express our sincere thanks to the symposia organizers, session chairs, presenters and conference attendees, for their efforts and enthusiastic participation in the vibrant and cutting-edge conference.

DILEEP SINGH and JONATHAN SALEM
Volume Editors

OXIDES FOR HIGH TEMPERATURE VIBRATION DAMPING OF TURBINE COATINGS

David R. Clarke
School of Engineering and Applied Sciences
Harvard University, Cambridge, MA 02138

ABSTRACT

The mechanical damping behavior of several oxides in the kHz regime is compared. Over the temperature range from room temperature to 1000°C, three oxides with a high concentration of oxygen vacancies exhibit pronounced damping at intermediate temperatures. Based on the results and a simple analysis, it is concluded that a search criteria for identifying oxides with high damping is that they are defect crystal structures and have both high ionic conductivity and low thermal conductivity. An expression for the peak damping temperature is given.

INTRODUCTION

Oxide coatings are currently being applied to high-temperature turbine blades to provide a thermal barrier enabling the gas temperatures of turbines to be increased without raising the temperature of the surface of the metal blades. This has allowed dramatic increases in turbine efficiency and power in the last decade, a period over which the development of higher temperature creep resistant single crystal superalloys has reached a level of maturity [1, 2]. This success raises the possibility that oxide coatings may provide additional functionality other than thermal resistance alone. In this contribution we describe recent exploratory work to investigate the potential of oxides to provide vibrational damping. Measurements at kHz frequencies are emphasized since these correspond to flexural vibrations induced in blades as a result of buffeting as they successfully pass behind vanes and into the turbine gas flow.

As a starting point for explorations we have paid most attention to oxides that have very low thermal conductivity so that the primary selection criterion for selecting the oxide as a thermal barrier is not compromised. In addition to the current coating material, yttria stabilized zirconia (YSZ), several different oxides with low thermal conductivity have been identified in the last decade, as shown in figure 1. It is emphasized that the data shown is the intrinsic conductivity, measured for fully dense materials, and that the introduction of porosity and gaps during processing can dramatically reduce the conductivity as is practiced in current coating technologies.

EXPERIMENTAL DETAILS

The vibration system we have used to measure flexural damping up to ~ 1200°C has been fully described elsewhere [3, 4]. Briefly, a cantilever beam of the material, clamped at one end, is set into vibration by broad-band excitation and the vibrational spectrum of displacements of the free end is monitored by a non-contact laser vibrometer. The dimensions of the beam are chosen so that the first resonance mode is at kHz frequencies. The resonant frequency is measured and the spectral analysis software is used to determine the damping ratio from the width of the resonance. These measurements are repeated at different temperatures and both the elastic modulus and damping ratios are calculated as a function of temperature. Although the vibration measurements are very rapid, the physical size of the beams together with the necessity of having the entire clamping and vibration system inside the furnace, to ensure constant temperature dictates that the overall system has a slow response time to changing temperatures, limiting the number of data points that can be recorded in a reasonable experimental period.

The oxides investigated were prepared from either oxide powders or powders prepared by standard solution routes. The powders were then compacted into discs and sintered to close to full density. The sintered disks were then cut to the shape of rectangular bars using diamond impregnated blades and subsequently diamond polished. Good surface finish is necessary to ensure high reflectivity of the laser beam used to monitor the beam displacements. For some samples, that were particularly transparent, a thin platinum metal film was deposited to increase the laser reflectivity. Careful grinding and polishing to ensure parallel sides of the beams is also necessary. Although fully dense oxides are not really required to make valid damping measurements, removal of flaws, particularly edge flaws, is desirable to ensure a firm grip in clamping the beam without breaking the sample while at the same time minimizing spurious vibrations and displacements. An experimental protocol was developed to ensure that the results were repeatable and reproducible.

RESULTS

An illustration of the damping of a typical TBC system, consisting of a 140 micron thick 7.6 mole % yttria-stabilized zirconia (7YSZ) coated PWA-1484 single crystal superalloy with an intermediary aluminide bond-coat, is shown in figure 2 together with the damping of an uncoated alloy. The comparison indicates that the 7YSZ coating produces damping at intermediate temperatures that is absent in either the bare superalloy or the aluminide coated superalloy. At the higher temperatures, above ~ 900°C there is a clearly defined damping peak superimposed on an increasing background. The higher temperature peak has been attributed to Ni and Al diffusional hopping in the □' phase of the superalloy but until our recent work the damping properties of this higher temperature regime had not been investigated [3]. The peak due to the YSZ occurs at a similar temperature as has previously been reported from low frequency internal friction measurements of single crystal zirconia [5]. In these other works the damping has been attributed to oxygen vacancy diffusional hopping. Using polycrystalline YSZ beams, we have shown that there is also an increasing damping background with increasing temperature above about 950°C whose identity has not been determined. [4]

The damping behavior of the four oxides, alumina (Al_2O_3), 7YSZ, gadolinium zirconate ($Gd_2Zr_2O_7$) and yttrium zirconate delta-phase ($Y_4Zr_3O_{12}$), have been investigated to date. It has been found that alumina does not exhibit any significant damping in the kHz range up to at least 1100°C whereas the other oxides do as is shown in figure 3. Furthermore, their damping peaks occur at significantly different temperatures, albeit at moderate temperatures in the range of 200-500°C.

DISCUSSION

In contrast to metals, comparatively little is known about the damping mechanisms in oxides, the magnitude of the damping attainable at different frequencies or their temperature dependence. In metals, damping can originate from a variety of defect mechanisms. For instance, the motion of interfaces, such as glissile twin interfaces in thermoelastic martensites or magnetostrictive alloys, from diffusional processes, such as Ni and Al vacancy diffusion in gamma-prime Ni_3Al, and from dislocation climb. These typically occur at relatively low frequencies. At very high frequencies, at MHz to GHz frequencies, energy can be dissipated by anharmonic phonon interactions, the so-called Akhiezer damping. At intermediate frequencies, thermoelastic damping can operate in all classes of material to a lesser or greater extent. This form of damping is associated with the generation of heat through the thermoelastic effect and the time constant for its flow from the hotter parts to the cooler parts of a solid during a vibrational cycle. The extent of damping depends on only on the material and its thermal diffusivity but also on the stress state. In the simplest case, that of a purely compressional vibration wave, heat is generated in the regions in compression and cooling occurs in the expanded part of the wave, so heat flows from the compressional to tensile regions of the material.

In the ceramics literature, damping in simple oxides such as rutile (TiO_2) and of thoria (ThO_2) containing CaO [6] and ZrO_2 has been identified with point defect reorientation under vibrational stresses. These classic studies, performed by standard internal friction measurements, have indicated that damping can occur by either cation interstitials (in rutile) or oxygen vacancies (in thoria), and give rise to damping peaks at relatively low temperatures, less than about 250°C. Since those early studies, almost all the measurements have since been on stabilized zirconia. These studies, which have been at low frequencies [7], have generally been performed to elucidate mechanisms of defect diffusion and its relation to ionic conductivity rather than of damping *per-se*.

The results shown in figure 3 clearly indicate that all three of the oxides measured to date other than alumina exhibit significant damping at intermediate temperatures. Furthermore, although damping is believed to originate from similar defect mechanisms there is a clear difference in the peak temperature. A common feature of the three oxides not shared by alumina is that they are all defective oxides, containing a high concentration of oxygen vacancies. In the case of the YSZ, the vacancies are associated with the concentration of Y^{3+} ions used to stabilize the YSZ. In both the $Gd_2Zr_2O_7$ and $Y_4Zr_3O_{12}$ compounds, the vacancies are intrinsic to their crystal structure. All three of the oxides have both exceptionally low and almost temperature-independent thermal conductivity [8], which is attributed to extensive phonon scattering as well as a mean free path being governed by the spacing of the vacancies. Based on these considerations, it is considered that the vibrational damping we observe is also associated with the vacancies in these compounds. If this is indeed correct, then this provides some guidance as to other possible oxide compounds that will also exhibit damping. There is some basis for this conclusion as discussed in the following paragraph.

At the present time, it is not possible to predict *a-priori* the peak damping temperatures for point defect damping of different compounds. However, some guidance comes from the analysis of defect relaxation originally presented by Wachtman [6]. He considered the rearrangement of point defects that could adopt a variety of equivalent crystallographic sites and how an alternating field affects the number on each site as they respond to the field. In effect, the point defect configurations represent different defect dipoles and in response to the applied field direction, the point dipoles reorient by the point defects diffusing to equivalent configurations but with lower energy. The energy dissipation is then controlled by the vibrational frequency applied and the diffusional jump rate. The latter, which depends on the energy barrier between the different defect sites, is related to temperature by an activation energy, E_i, similar to that for diffusion. The relationship that Wachtman derives between the damping factor, Q^{-1}, temperature, T, frequency, ω , and relaxation time, τ , can be expressed as

$$Q^{-1} = \frac{A\,Y\,N}{k_B T}\,\mathbf{sech}\left[\ln(\omega\,\tau) + {E_i}/{k_B T}\right] + \frac{constant}{T} \tag{1}$$

where Y is Young's modulus, N is the concentration of defects, k_B , is Boltzmann's constant and A is a constant that depends on the material. The damping factor then depends on temperature in a number of distinct ways, explicitly in the reciprocal in the first term and through the activation energy in the relaxation rate. As can be judged visually from the fit to our data in figure 3, the form of equation 1 captures the functional dependence on temperature exhibited by the three oxides. The equation also provides another insight, namely the dependence of the damping peak temperature, T_P. From examination of the form of equation 1, the damping peak occurs when the term inside the *sech* has a zero value, namely

$$T_P = \frac{E_i}{k_B\,|\ln \omega\,\tau|} \tag{2}$$

Although Wachtman's analysis is based on a dilute concentration of defects whereas the concentration of defects in the oxides we have studied is of the order of tens of percent, it has a number of dependencies that can be correlated with other properties. For instance, the dependence of damping on a high concentration of defects is consistent with the requirement for producing low thermal conductivity by phonon-defect scattering. Similarly, the dependence on low activation energy for diffusional hopping from one site to another is consistent with an oxide having high ionic mobility and hence conductivity. More rigorous evaluation on a larger number of oxides is needed before Wachtman's analysis can be demonstrated to be a robust guide to other oxides but it does support the conclusion that high damping factors can be expected from defective oxides.

Finally, although the foregoing discussion has been concerned with damping associated with point defects, it does not rule out the possibility that other damping mechanisms, such as ferroelastic twinning, might operate in other oxides. Nevertheless, it does suggest that high ionic conductivity and low thermal conductivity will be useful criteria in identifying candidate oxides for damping applications.

CONCLUDING REMARKS

The results presented in this work suggest that defective oxides may hold promise as coating materials combining the functionalities of serving as a thermal barrier coatings and damping elements in future turbine blade designs. The coatings investigated to date exhibit damping at temperatures more appropriate to certain compressor turbines stages and hence have the potential of being implemented to complement existing, mechanical damping schemes. Furthermore, the observation that the damping peak temperature varies from oxide to oxide, suggests that there is scope in using existing crystal chemistry rules together with othe property correlations for identifying oxides that could exhibit damping at still higher temperatures.

ACKNOWLEDGEMENTS

This work was supported by the Office of Naval Research under program grant N000-04-1-0053.

REFERENCES

[1]"Coatings for High Temperature Structural Materials: Trends and Opportunities," National Research Council, Washington, 1996.

[2]U. Schulz, et al., "Some Recent Trends in Research and Technology of Advanced Thermal Barrier Coatings," *Aerospace Science and Technology*, 7 73-80 (2003).

[3]G. Gregori, et al., "Vibration Damping of Superalloys and Thermal Barrier Coatings at High Temperatures," *Materials Science and Engineering, A*, 466 256-264 (2007).

[4]A.M. Limarga, et al., "High-Temperature Vibration Damping of Thermal Barrier Coating Materials," *Surface and Coating Technology*, 202 693-697 (2007).

[5]M. Weller, et al., "Oxygen Mobility in Yttria-Doped Zirconia Studied by Internal Friction, Electrical Conductivity and Tracer Diffusion Experiments," *Sold State Ionics*, 175 409-413 (2004).

[6]J.B. Wachtman, "Mechanical and Electrical Relaxation in ThO_2 Containing CaO," *Physical Review*, 131 517 (1963).

[7]M. Weller, B. Damson, and A. Lakki, "Mechanical Loss of Cubic Zirconia," *Journal of Alloys and Compounds*, 310 47-53 (2000).

[8]M.R. Winter and D.R. Clarke, "Oxide Materials with Low Thermal Conductivity," *Journal of the American Ceramic Society*, 90[2] 533-540 (2007).

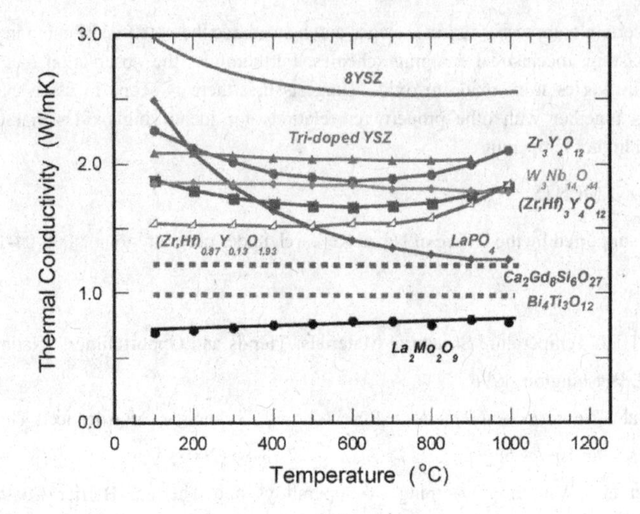

Fig 1. Thermal conductivity of a variety of oxides as a function of temperature. The data, obtained by thermal flash measurements, is for fully dense oxides.

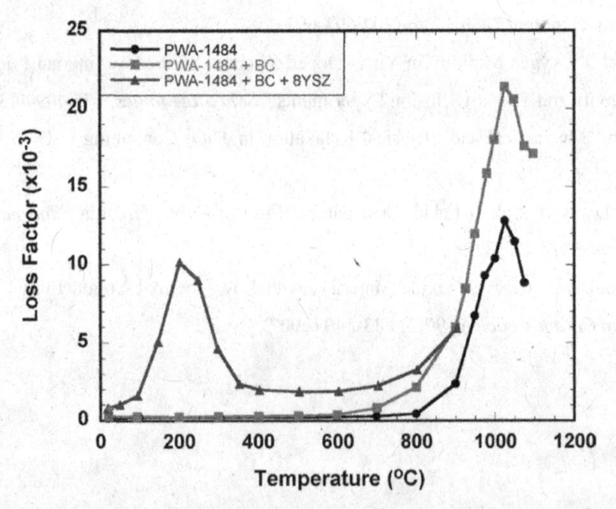

Fig 2. First resonance flexural vibration mode damping as a function of temperature for a thermal barrier coating system as well as data for the PWA 1484 superalloy together with an aluminide bond coated (BC) superalloy .

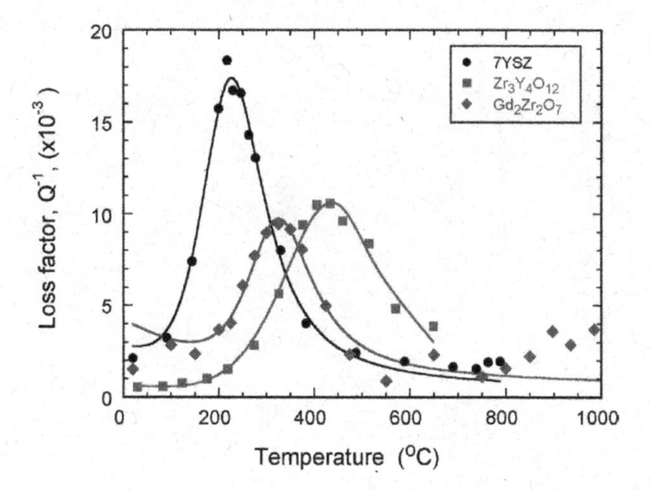

Fig 3. Comparison of the temperature dependent flexural damping of three defective oxides in the kHz frequency range. The curves through the data are the best fit to the functional form of equation 1 in the text.

ENHANCING THE PASSIVE DAMPING OF PLASMA SPRAYED CERAMIC COATINGS

J. P. Henderson, A. D. Nashif, J. E. Hansel, (Universal Technology Corporation, Dayton, OH) and R. M. Willson (APS Materials Inc., Dayton, OH)

ABSTRACT

Hard coatings deposited by plasma spray can dissipate vibratory energy as well as enhance the ability of turbine components to withstand the environment. However, with no other dissipative mechanism, reducing system quality factors (Q) in bending vibration to a target of 100 while increasing thickness no more than 10% requires a material loss modulus of at least 3.7 GPa (0.54 Mpsi) at strains of interest. Such ceramics as magnesium-aluminate spinel, alumina, titania-alumina, and yttria stabilized zirconia have been found to have loss moduli of only 1 -2 GPa. Preliminary work has shown that vacuum infiltrating alumina with a viscoelastic material (VEM) chosen for effectiveness at about 200F (93C) increased the loss modulus at strains of 400 ppm by a factors of 2 at room temperature and 3 at the design temperature. Infiltration of plasma-sprayed titania-alumina with the same VEM showed increases of factors of 3 and 4, respectively. An infiltrate with a higher glass transition temperature enables high damping at higher temperatures. In current work, particles of high temperature viscoelastic material (HTVEM) are co-sprayed with yttria stabilized zirconia. Preliminary results at low levels of strain suggest that the loss modulus of such materials at 1000-1400F (540-760C) may be as much as four times that obtained with the low temperature infiltrate at 93C.

INTRODUCTION

The requirement for increasing damping of components in modern turbine engines, particularly those used in aircraft, has been a subject of special emphasis in the National Turbine High Cycle Fatigue (HCF) Program that was initiated in 1994. Some of the drivers for the increased interest in damping include the increased utilization of Integrally Bladed Rotors (IBRs), or BLISKS, and high stage loadings with lower aspect ratio blades and vanes. IBR structures, in which the blades and rotor are a single piece, have less inherent damping than rotors with inserted blades that dissipated some energy, due to friction, in the dove-tail interfaces. Lower aspect ratio airfoils have resulted in higher modal densities of high order, high frequency modes, making the task of avoiding the excitation of these modes much more difficult. At some operating conditions a frequency of excitation from aerodynamic wakes can coincide with a resonant frequency of a blade. Once a resonant mode is excited and sustained the only thing limiting the maximum stress in the blade is damping.

Several approaches to increasing the damping of turbine engine components have been investigated. Some of these approaches include better modelling of frictional damping to improve designs of blade shrouds, platforms, and wedge dampers between blades; internal stick dampers inside blades, air film dampers, hollow blades filled with viscoelastic materials, embedded viscoelastic constrained layers, and coatings. None of these approaches are a universal panacea that will solve all resonant response problems that present a risk of HCF failures. Damping coatings, however, show promise of being one of the least invasive methods for increasing damping of specific components in a turbine engine. It is important to remember that turbine engine damping coatings must do much more than damp resonant vibrations. To be effective these coatings must be able to survive the severe creep, erosion, impact and fatigue environments associated with turbine engines.

BACKGROUND

The properties of the damping coatings investigated in this program have been modeled in terms of complex Young's modulus, E^*. Where

$$E^* = E_1 + jE_2 = E_1 (1+j\eta)$$

and, by definition,

E_1 = Young's Storage Modulus

E_2 = Young's Loss Modulus

$$\eta = \text{Loss Factor} = E_2/E_1 \quad (1)$$

The coatings are considered to be free-layer damping treatments, analyzed by Öberst [1] in 1952, consisting of a homogenous layer with complex modulus attached to an elastic beam. A free-layer coating on a single side of a beam is illustrated in Figure 1.[2]

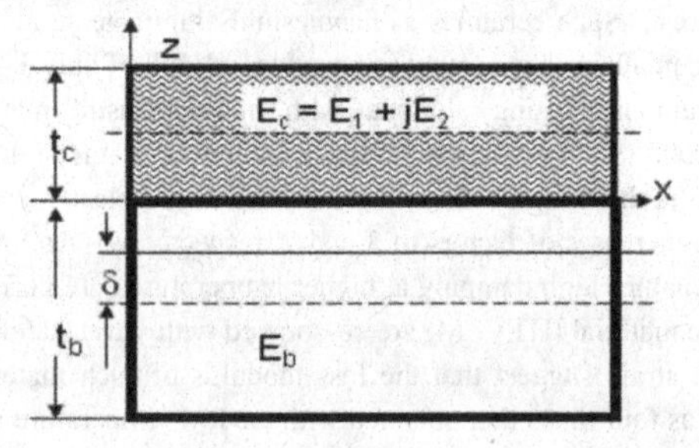

Figure 1. A Free-Layer Damping Coating on a Beam[2]

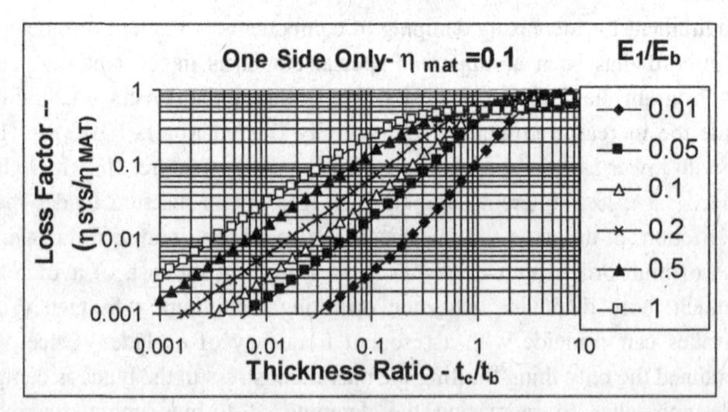

Figure 2. Normalized System Loss Factors (1/Q) of a Beam with a Free-Layer Damping Coating[2]

Plasma spayed ceramic coatings are not strictly homogeneous layers, as they typically consist of a thin bond coat with a thicker top coat, but characterization of the combined bond coat/top coat system as a single free-layer damping coating gives valid engineering data. If the usual simplifying assumptions are made that the coating is thin compared to the beam and that shear deformations can be neglected, damping performance is proportional to the thickness ratio (t_c/t_b) and the loss modulus of the free layer coating, as illustrated in Figure 2[2]. For a given thickness of a damping coating, the one measure of damping merit is loss modulus.

POLYMERIC FREE-LAYER DAMPING COATINGS

The original free-layer damping treatments, analyzed by Öberst, were polymeric viscoelastic damping layers. In the early 1960's Monsanto and Lord Manufacturing Company collaborated in marketing a free-layer damping treatment called LD-400. This material was developed and sold primarily for the reduction of noise radiating from heavy steel structures undergoing resonant response in thermal environments close to room temperature. LD-400 was an exceptional free layer damping treatment due to its high loss modulus which was achieved by filling the co-polymer matrix with flake graphite and orienting the flakes parallel to the surface of the material[3]. As with all viscoelastic materials, the damping properties of LD-400, expressed in terms of complex modulus, are functions of both temperature and frequency. Figure 3 shows damping properties of LD-400 at 1000 Hz, as developed from measurements.

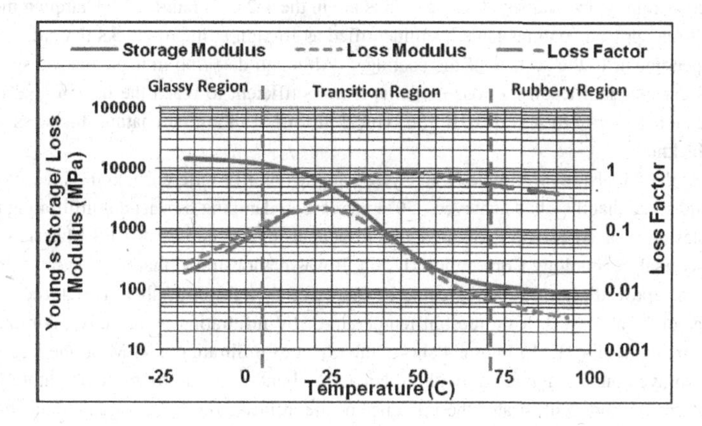

Figure 3. Damping Properties of LD-400 at 1000 Hz.

The properties shown in Figure 3 exemplify the behavior of viscoelastic polymers. At cold temperatures (<10C for LD-400) the storage modulus is high and the loss factor and loss modulus are low. This temperature region is referred to as the "glassy" region. As temperature increases the storage modulus drops rapidly as a function of temperature in the "transition" temperature region (>10C <70C for LD-400) and the loss factor peaks. The loss modulus peaks in the lower part of the "transition" temperature region. At higher temperatures the storage modulus levels off accompanied by a reduced loss factor and loss modulus in the "rubbery" region (>70 C for LD-400). It can be seen from Figure 3 that LD-400 exhibits excellent damping properties at 1000 Hz near room temperature, and has a loss modulus > 1000 MPa in the temperature range >5C and < 40C. LD-400 remains an example of one of the most effective polymeric free-layer damping treatments.

Although free-layer polymeric damping treatments have been successfully applied in industrial noise control of heavy steel structures, they are not appropriate for most aerospace applications. Free-layer polymeric damping treatments are too heavy, too thick and too sensitive to creep and erosion deterioration to be seriously considered for use in most aircraft turbine engines.

PLASMA SPRAYED CERAMIC DAMPING COATINGS

One class of free-layer damping treatments that has been considered for use in turbine engines is plasma sprayed ceramic coatings. Such ceramics as magnesium-aluminate spinel, alumina, titania-alumina, and yttria stabilized zirconia have been found to have significant damping.[4] Unlike viscoelastic materials (VEM), these hard ceramic coatings are not very sensitive to changes in temperature or frequency, but they are non-linear with respect to dynamic strain, i.e., the material properties are dependent on amplitude of strain. At very low vibratory strains these plasma sprayed coatings exhibit insignificant values of loss modulus. As dynamic strain increases these coatings show a reduction in storage modulus and an increase loss modulus. At dynamic strains of approximately 400 micro-strain typical values of loss modulus are in the 1-2 GPa range.[4, 5, 6] Damping mechanisms in these plasma sprayed coatings have been identified as friction at micro-cracks that are both parallel to and perpendicular to the surface of the coating.[5] Although damping in these plasma sprayed ceramic coatings can be significant, the loss modulus is not sufficient to reach the design goal of a resonant amplification factor Q of less than 100 in typical airfoils with a total coating thickness < 10% of the airfoil thickness.

At the 8th National Turbine High Cycle Fatigue (HCF) Conference, 2003, investigators at Rolls Royce indicated that they had achieved >50% increase in damping by vacuum infiltrating polyurethane into a plasma sprayed ceramic coating.[5] In subsequent investigations by APS Materials Inc., teamed with Universal Technology Corporation (UTC), it was demonstrated that the damping (loss modulus) of a plasma sprayed ceramic coating could be increased by a factor of about two at room temperature[6] and a factor of at least 3 at the design temperature[7] by infiltrating a carefully optimized polymeric viscoelastic material (VEM). Figure 4 shows the effect of infiltrating a VEM on the loss modulus of a plasma sprayed alumina coating with a NiCrAlY bond coat at room temperature and at high temperatures. Figure 5 illustrates the influence of the infiltrate on the storage modulus of the alumina coating. It is important to note that the plasma sprayed ceramic coatings investigated by APS Materials Inc. were not porous coatings. These coatings had densities >95% and the amount of infiltrated VEM was very small. As the damping of the infiltrated coatings is temperature sensitive, specific VEMs were selected to match the thermal environments of the components operating in the cold section of the engine. Plasma sprayed ceramic coatings, blends of alumina and titania, were infiltrated with a VEM optimized for application on specific turbine engine fan blades. They were tested in a spin-pit to centrifugal loads of 80,000 G at temperatures up to 200C with no indication of creep. Tests on these coatings showed no significant reduction in erosion resistance when compared to that of the bare blade.

High cycle fatigue tests showed a small reduction in HCF fatigue strength of the coated titanium alloy, but the reduction in strength is inconsequential when compared to the reduction in stress resulting from the application of the damping coating.

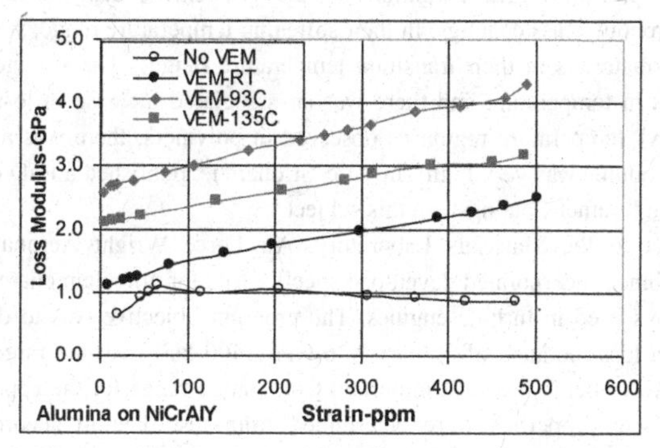

Figure 4. Influence of VEM Infiltrate on Young's Loss Modulus of an Alumina Coating [6,7]

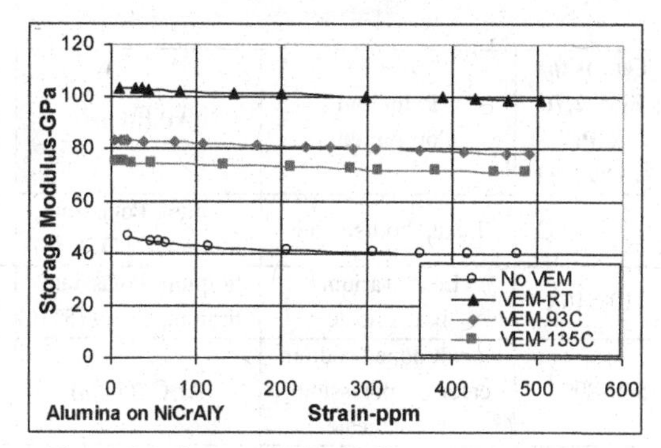

Figure 5. Influence of VEM Infiltrate Young's Storage Modulus of an Alumina Coating[6,7]

The ability to predict damping of a coated fan blade utilizing the measured strain-dependent complex modulus has been demonstrated with excellent correlation between analytical and experimental results.[8]

The actual temperature limit for a VEM infiltrated plasma sprayed coating depends not only on the temperature, but also on the time at temperature.[9] Several polymeric VEMs have been identified as candidates for infiltration of plasma sprayed ceramic coatings operating at temperatures generally <300C.

HIGH TEMPERATURE VISCOELASTIC DAMPING MATERIALS (HTVEM)

There has been a long term desire to find effective damping coatings for turbine engine components operating at temperatures well above 300C. In the 1960s B. J. Lazan, at the University of Minnesota, did some preliminary investigations of the viscoelastic behavior of porcelain enamel coatings. These amorphous glass coatings, in their softening temperature range, were found to behave much like polymeric materials in their transition temperature range. The storage modulus dropped rapidly with increases in temperature and there was an associated increase in loss factor. Although there was no "rubbery" temperature region as observed in polymers, there was a broad temperature range in which loss modulus was very high. In 1975 Sridharan[10] published a PhD dissertation entitled "Damping in Porcelain Enamel Coatings" on this subject.

Under contract to the Materials Laboratory, Air Force Wright Aeronautical Laboratories (AFWAL), Solar Turbines Incorporated developed specific frits for porcelain enamels to be applied to high temperature alloys used in turbine engines. The program objective was to develop a family of enamels to provide a loss modulus of at least 2.76 GPa (400 ksi) over the range from 400-870 C, identified by the USAF as being the most important temperature range for the application of damping treatments.[11] Damping properties were determined from test data in accord with the ASTM standard.[12] A family of five frits, selected from the 16 frits considered, were found to meet the program objective, with a small gap in coverage around 540 C. Selected attributes of these five are shown in Table I.

Material Code	Temps for $E_2 \geq 2.76$ GPa @1000 Hz	Principal Components	Additions	Density (gm/cc)
S1-B	370-490 C	Lead, borosilicate	Sodium, Potassium, lithium, Cr_2O_3 (5%)	4.8
S3-B	410-510 C	Lead, barium, borosilicate	Sodium, Potassium, lithium, Cr_2O_3 (5%)	3.61
S5-D	560-800 C	Cobalt doped sodium, calcium, potassium, borosilicate	Al_2O_3 (17%)	2.65
S6-B	580-825 C	Complex barium, borosilicate	Cr_2O_3 (5%)	3.01
S13-B	760-900 C	Barium, calcium, borosilicate	Cr_2O_3 (5%)	3.17

Table I. Attributes of Damping Coatings by Solar Turbines Incorporated[11]

Even though these porcelain enamels had outstanding damping properties due to their high loss modulus values in their optimum temperature ranges,[11] they have been only utilized on only a few static engine components.[13] The major problems that limited the application of these coatings in turbine engines were due to creep, especially on rotating blades and erosion due to high velocity airflow.

PLASMA SPRAYED CERAMICS SMALL INCLUSIONS OF HTVEM

Encouraged by the success of infiltrating small quantities of polymeric VEM in plasma sprayed ceramic coatings it was decided to attempt to introduce small quantities of HTVEM in a plasma sprayed yttria stabilized zirconia (YSZ) coating. While many enamels appear to have the potential for high damping above about 650C, identification of enamels with the desired properties at somewhat lower temperatures proved to be more difficult. Several candidates were identified on the basis of softening temperature and other factors. Test specimens were then prepared by mixing the YSZ powders with the frits and co-spraying the mixtures onto Hastelloy X test specimens using a NiCrAlY bond coat. Testing for this effort was conducted using the forced response approach previously used for the polymeric infiltrate modified to acquire data at higher temperatures.[6] Specimen geometry was modified from that used in the previous work through the incorporation of a long root, as shown in Figure 6, to enable insertion of the test section into a high temperature oven. With this long specimen root, the fixture for holding the cantilever specimen was maintained at a reasonably low temperature to reduce the damping in the clamp. Only two inches of the specimen root section entered the clamping blocks, as shown by the stippled regions in Figure 6. Temperature variation along the length of the test section was held to <55C when testing at 800C.

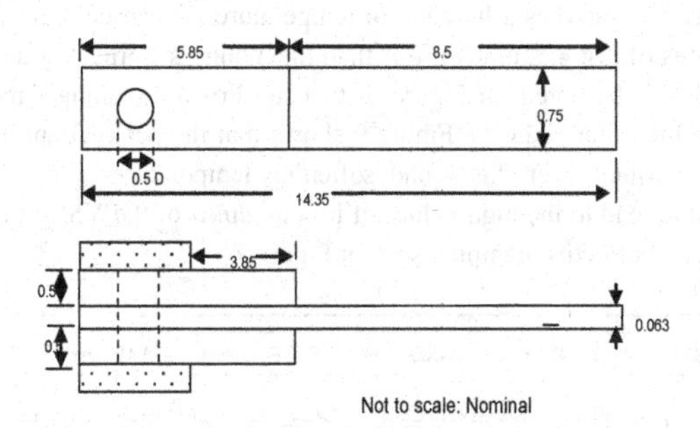

Figure 6: Specimen used for High Temperature Damping Tests

Resonant frequencies and modal damping were measured for both bare beams and coated beams vibrating in the second through the forth bending modes of the test section. As the coatings with HTVEM were found to show much less amplitude dependence than those of uninfiltrated coatings or of coatings infiltrated with VEM, the determinations of material properties were made by the methodology appropriate for linear viscoelastic behavior.[12] A comparison of the modal responses these specimens, along with information on the geometries and densities of the beams and coatings, was used to calculate nomograms relating temperature and frequency dependence of the coatings to be displayed on a single master curve.[14, 15]

One material, identified as 'Frit 4' was selected as having the high damping over the broadest range of temperatures. Figure 7 shows the measured loss modulus of the YSZ coating with different weight % loading of Frit 4 at 1000Hz. The loss modulus of YSZ alone at low strain, as determined by the same methodology, is shown for reference.

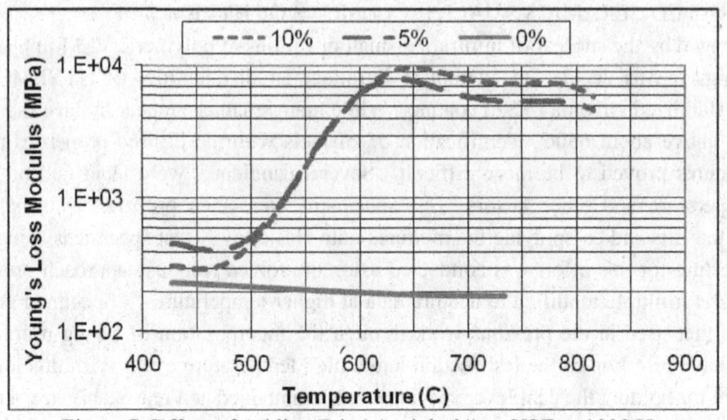

Figure 7. Effect of Adding Frit 4 (weight %) to YSZ at 1000 Hz

Frit 4 was also applied to a test beam as a homogeneous enamel free-layer coating. Properties of this enamel coating were measured as a function of temperature. It can be seen from Figure 8 that the Young's storage modulus of Frit 4 is much lower than the Young's storage modulus values for the Solar enamels.[11] It should also be noted, in Figure 8, that the Frit 4 softening temperature range is much wider than it is with the Solar enamels. Figure 9 shows that the peak Young's loss modulus of Frit 4 remains relatively constant over this broad softening temperature range. It is these Frit 4 properties that are believed to lead to the high values of loss modulus of the YSZ/Frit 4 composite and the broad temperature range of effective damping seen in Figure 7.

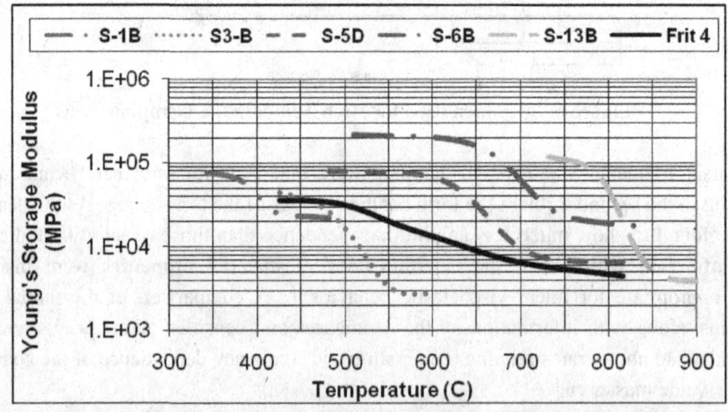

Figure 8. Young's Storage Modulus of Frit 4 Compared with Solar Enamels at 1000 Hz

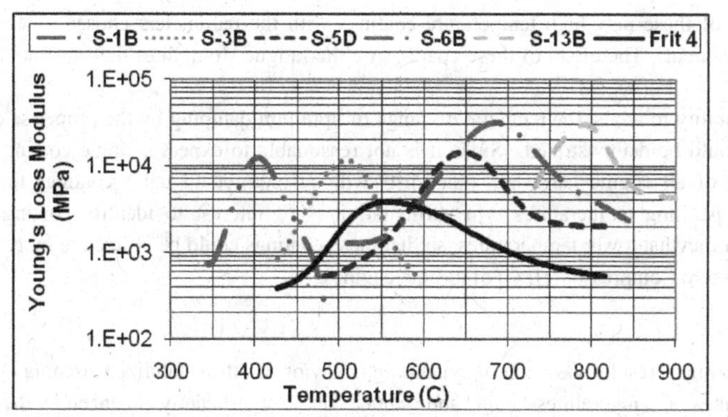

Figure 9. Young's Loss Modulus of Frit 4 Compared with Solar Enamels at 1000 Hz

It is interesting to compare, in Figure 10 the loss modulus/temperature plots for LD-400 (an exceptional polymeric free-layer damping treatment), a plasma spayed titania blend coating infiltrated with APS 600 VEM (a coating that has demonstrated capability in the damping of a fan blade), Solar S-6B (a porcelain enamel with outstanding damping but little resistance to creep and erosion), and a YSZ coating with an addition of 10% Frit 4. The YSZ/Frit 4 coating at 650C exhibits four times the loss modulus of the lower temperature VEM infiltrated ceramic coating at its optimum temperature of 200C. This means that at their optimum temperatures the YSZ/Frit 4 coating could provide twice the damping with a coating one-half as thick as the lower temperature coating.

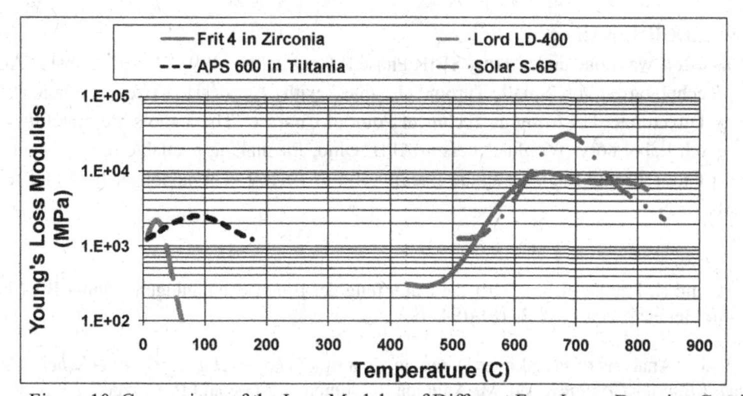

Figure 10. Comparison of the Loss Modulus of Different Free-Layer Damping Coatings

The titania blend coating infiltrated with APS 600 demonstrated considerable creep resistance at temperatures well above the transition temperature range, as well as erosion resistance. But the Solar S-6B material, while found to be an outstanding damping coating, lacks the creep and erosion resistance required for a coating to be used on rotating airfoils. Clearly there is a need for creep and erosion tests of these new high temperature coatings with the parameters chosen as appropriate for specific components. The effect of these coatings on the fatigue strength of the substrate must also be evaluated.

The ability to adjust the temperature range of optimum damping by the proper selection of the HTVEM, should be demonstrated. Since it is not reasonable to expect a single coating to meet the requirements of all engine hardware, other frits will be required to tailor coatings to the specific component operating temperatures. In particular, it is of interest to identify coatings with peak damping at somewhat lower temperatures, so that these coatings could be utilized in more situations in the High Pressure Compressor (HPC) of turbine engines.

SUMMARY

Preliminary results show that the damping behavior of yttria stabilized zirconia (YSZ) plasma sprayed coatings, at temperatures >550C and <800C, can be significantly enhanced by the inclusion of small quantities of High Temperature Viscoelastic Materials (HTVEM). The resulting modified coatings have excellent damping as indicated by comparing measured loss moduli with those of other quality free-layer damping treatments. These measurements indicate the potential of utilizing plasma sprayed damping coatings in turbine engines at temperatures well above the useable range of polymeric viscoelastic material (VEM) infiltrated plasma sprayed ceramic damping coatings. Although these results are very encouraging, additional investigations are required to determine the suitability of these coatings for turbine engine applications. In particular, the resistance to erosion and the impact on substrate fatigue must be investigated. And finally, it must be established that creep migration of the HTVEM at high temperatures does not negate the high damping in a 'high-g' rotational environment.

ACKNOWLEDGEMENTS

This effort was done under Army SBIR Phase II Contract No. W911W6-07-C-0043 "Advanced Damping Technologies for Small Turbine Engines" with Anastasia Kozup, Aviation Applied Technology Directorate, US Army as Technical Point of Contact. The authors are grateful to the Air Force Research Laboratory, Wright-Patterson AFB, Ohio, for making available the test facilities and equipment of the Turbine Engine Fatigue Facility and to Peter J. Torvik for analysis and review of these results.

REFERENCES

[1] Öberst, H. and K. Frankenfield, "Uber die Dampfung der Biegeschwingungen dunner Blesche durch festhafttende Belage," *Acustica,* **2**, 181-194, 1952,

[2] Torvik, P. J. "Analysis of Free Layer Damping Coatings," *Layered, Functional Gradient Ceramics, and Thermal Barrier Coatings*, Ed. M. Anglada, E. Jiménez-Piqué and P. Hvizdoš, *Key Engineering Materials,* Vol. 333, pp. 195-214, 2007.

[3] Ball, G.L. and I. O. Salyer,"Development of a Viscoelastic Composition Having Superior Vibration Damping Capability," J. Acoust. Soc. Am. 39:663-72, 1966.

[4] Torvik, P. J., "A Survey of the Damping Properties of Hard Coatings for Turbine Engine Blades, *Integration of Machinery Failure Prevention Technologies into System Health Management*, Society for Machine Failure Prevention Technology (MFPT), Dayton, OH, pp 485-506, 2007.

[5] Shipton, M. H., and S. Patsias, "Hard Damping Coatings: Internal friction as the damping mechanism," *Proceedings of the 8th National Turbine High Cycle Fatigue (HCF) Conference*, Monterey, CA, 14-16 April 2003.

[6] Torvik, P., R. Willson, and J. Hansel, "Influence of a Viscoelastic Surface Infiltrate on the Damping Properties of Plasma Sprayed Alumina Coatings Part I: Room Temperature," *Proceedings, Materials and Science & Technology 2007 Conference and Exposition*, Detroit, MI, Sept 16-20, 2007.

[7] Torvik, P., R. Willson, J. Hansel, and J. Henderson, "Influence of a Viscoelastic Surface Infiltrate on the Damping Properties of Plasma Sprayed Alumina Coatings Part II: Effects of elevated Temperature and Static Strain," *Proceedings, Materials and Science & Technology 2007 Conference and Exposition*, Detroit, MI, Sept 16-20, 2007.

[8] Filippi, S and P. J. Torvik, "A Methodology for Predicting the Response of Airfoils with Non-linear Coatings," (in preparation for publication).

[9] Nashif,A. D., "How to Predict the Effects of Aging On the Dynamic Properties of Viscoelastic Materials", *Proceedings of the 75th Shock and Vibration Symposium*, Virginia Beach, VA, Oct 17-22, 2004.

[10] Sridharan, R. I., "Damping in Porcelain Enamel Coatings, AFWAL-TR-74-191, Materials Laboratory, Wright-Patterson AFB, Ohio,1976 (also published as a PhD dissertation, University of Minnesota, 1975).

[11] Brentnall, W. D., A. R. Stetson, A. G. Metcalfe, and A. D. Nashif, "Enamels for Engine Structural Damping", AFWAL-TR-83-4110, Materials Laboratory, Wight-Patterson AFB, Ohio, October 1983 (Cleared for public release ASC-00-2036, 4 Oct 2000).

[12] ASTM Subcommittee e33.03, *Standard Test Method for Measuring Vibration Damping Properties of Materials*, ASTM E 756-05, ASTM International, West Conshohocken, PA, 2005.]

[13] Jones, D. I.G., and C. M. Cannon, "Control of Gas Turbine Stator Blade Vibrations by Means of Enamel Coatings," *Journal of Aircraft*, Vol 12, No 4., pp. 226-230, April 1975.

[14] Nashif. A. D., "Measurements and Modeling of the Damping Properties of Viscoelastic Materials", Universal Technology Corp (UTC) Report, 2008.

[15] Nashif, A. D., D. I. G. Jones, and J. P. Henderson: *Vibration Damping,* Wiley Interscience, New York, pp 339-348, 1985.

MAGNESIA AND YTTRIA BASED COATINGS FOR DIRECT-COPPER-BONDING OF SILICON NITRIDE CERAMICS

L. Mueller[1], T. Frey[1], A. Roosen[2] and J. Schulz-Harder[3]

[1]University of Applied Sciences Nuremberg, Faculty of Materials Engineering, Wassertorstrasse 10, 90489 Nuremberg, Germany
[2]University Erlangen-Nuremberg, Department of Materials Science, Glass and Ceramics, Martensstrasse 5, 91058 Erlangen, Germany
[3]Electrovac curamik GmbH, Im Gewerbepark D 75, 93059 Regensburg, Germany
Corresponding author email: lars.mueller@ohm-hochschule.de

ABSTRACT

Besides excellent bending strength, high fracture toughness and good thermal conductivity Si_3N_4 ceramics are of great interest for DCB substrates in power electronic applications. Unlike oxide ceramics like Al_2O_3, which are commonly used as a DCB substrate, it is hard to produce DCB coated Si_3N_4 circuit boards due to poor bonding of Cu. Therefore this work investigates adherence mechanisms between copper and MgO or Y_2O_3 coated Si_3N_4 ceramics, respectively. Oxide layers of various composition and thickness were formed by sol gel technique via dip coating. The microstructure was characterized by grazing incidence XRD and SEM as a function of Si_3N_4 oxidation and annealing. The stability of the compounds and the reversibility of chemical reactions appear to be dependent on the amount of oxygen present in the system as well as on layer composition and structure. The bonding strength of the Cu metallization on the coated Si_3N_4 was determined by shear tests. The chemistry of the substrate strongly affects the interfacial bonding strength. Copper on MgO coated Si_3N_4 exhibited higher peel adhesion strength than on Y_2O_3 coated Si_3N_4. This could be attributed to the formation of different reaction products in the interface and a pore-free contact area.

INTRODUCTION

To reach the ambitious targets in the automotive industry concerning emission reduction one of the most promising approaches is the widespread use of hybrid vehicles. In this field of application the harsh thermal and mechanical stress situation in particular for full hybrid (30-60 kW) and power hybrid (> 60 kW) applications require new kinds of substrates with outstanding structural and thermal performance for high power dissipation to meet the given challenges [1]. In recent years direct bonded copper (DBC) alumina and aluminium nitride substrates have made a great contribution to the packaging of power electronics [2]. However, both ceramic materials have at least one limitation. Al_2O_3 substrates have their limit with respect to the heat dissipation efficiency and AlN substrates do not exhibit the required bending strength and fracture toughness. Due to these impairments, Si_3N_4 ceramic substrates are gaining increasing interest for DCB applications due to their high heat conductivity and excellent mechanical strength [3,4].

The use of DCB- Si_3N_4 substrates is inhibited by the lack of a reliable and cost-effective DCB-coating process. Using active metal brazing can cause embrittlement or debonding by interfacial reaction products [5]. In addition, the desired vacuum increases the production costs [6]. The less expensive gas-metal eutectic bonding via the DCB process, which has become more popular in industry over the past decades, was used in the nineties to develop DCB-Si_3N_4 substrates. Using Si_3N_4 ceramics of low thermal conductivity a bonding could be achieved via glass formation after pre-oxidizing of Y_2O_3+Al_2O_3-doped or MgO-doped Si_3N_4 ceramics at high temperature [7,8]. The microstructure and sintering additives of Si_3N_4 substrates of high thermal conductivity differ considerably from conventional Si_3N_4 ceramic substrates [4] and specifically their ceramic phase composition affects disadvantageously the oxidation behaviour, interfacial chemistry and strength of direct bonded copper [9,10]. To be more independent of additives in the Si_3N_4 substrate and to avoid blistering through reactions between liquid CuO/Cu_2O and Si_3N_4 grains efforts has been taken to coat the substrate with alumina before the DCB process was applied [11]. Nevertheless, no DCB silicon nitride substrates with high thermal conductivity are in use so far.

In this paper, the common sintering additives for Si_3N_4 like Y_2O_3 and MgO were chosen to form intermediate layers by a sol gel technique and dip-coating process on two different Si_3N_4 substrates of high thermal conductivity. After subsequent annealing, these coatings were compared with the microstructure of uncoated substrates after oxidation. On the coated and uncoated Si_3N_4 substrates the DCB process was performed and the adherence mechanisms between magnesia- and yttria-coated Si_3N_4 ceramics, respectively, and copper were investigated.

EXPERIMENTAL METHOD

The used Si_3N_4 substrates of supplier A and B had a size of 20 mm x 50 mm x 0.3 mm and a thermal conductivity of 80 ± 15 W/(m·K), according to the vendor. Table 1 summarizes the chemical composition, density and surface roughness of the substrates as measured by the authors. The copper foils, alloyed with silver, and of 0.3 mm thickness were degreased, cut into stripes of 10 mm width and thermally pre-oxidized in air for 40 min at 270 °C to produce a CuO/Cu_2O layer of approximately 500 nm in thickness [12]. All Si_3N_4 substrates were cleaned by ultrasonic treatment in acetone and subsequently in ethanol for 10 min before the coating and oxidizing process, respectively.

Table I. Properties of as-fired Si_3N_4 substrates

Si_3N_4 ceramics	Composition (wt. %)					Density (g/cm^3)	Surface roughness R_a (μm)
	O	Al	Hf	Mg	Y		
Supplier A (Y-SN)	3.10	0,02	1.55	0.50	<u>2.70</u>	3.57±0.05	0.72±0.04
Supplier B (Mg-SN)	1.72	0,01	0.02	<u>1.50</u>	0.85	3.15±0.05	0.56±0.05

For chemical solution deposition (CSD) the sol-gel technique in combination with dip-coating was used to form nano-sized, uniform MgO and Y_2O_3 particulate coatings on the as-fired Si_3N_4 surface and to reduce the sintering temperature below the oxidation temperature of β-Si_3N_4 grains. The Mg-precursor sol solutions were prepared by the reaction of Mg ethoxide in absolute EtOH using glacial acetic acid as a steric acid-catalyst to yield primarily linear polymers. A Mg alkoxide was chosen as a source material for the sol-gel synthesis over Mg acetate or nitrate in order to avoid the formation of premature MgOH precipitates, sintering-inhibiting phases and additional water, because of its low evaporation rate during dip-coating [13]. The yttrium hydroxide sol was produced via hydrolysis and polycondensation of yttrium (III) isopropoxide dissolved in absolute EtOH mixed with a small amount of acetic acid. The viscosity (RheoStress 6000, Thermo Haake GmbH, Karlsruhe, Germany) and the pH (Voltcraft, Conrad Electronic GmbH, Hirschau, Germany) of the sols was measured.

The manufacturing route of the DCB-Si_3N_4 substrates is shown in Fig. 1. Dip-coating was conducted by dipping the yttria and magnesia rich Si_3N_4 substrates into the coating solutions shown in Table II, followed by slow withdrawal, both under constant dry ambient conditions to control particle size as well as morphology of the hydrolysed products. The aim of this treatment was to cover the specimen surface with a coating of at least 250 nm thickness. Without any aging time, the coated substrates were heat treated in air by firing up to 500 °C (dwell time 1 h) using a heating rate of 120 °C/h, followed by a second firing step at 1300°C (dwell time 1 h) using a heating rate of 300 °C/h. All dip-coatings were produced using a double dip-coating and heating cycle. The copper stripes were bonded to all coated substrates by the bonding process known as copper-oxygen eutectic method as described elsewhere [14]. In general, the DCB process was performed on both sides of the tape cast substrates, which differ in their microstructure.

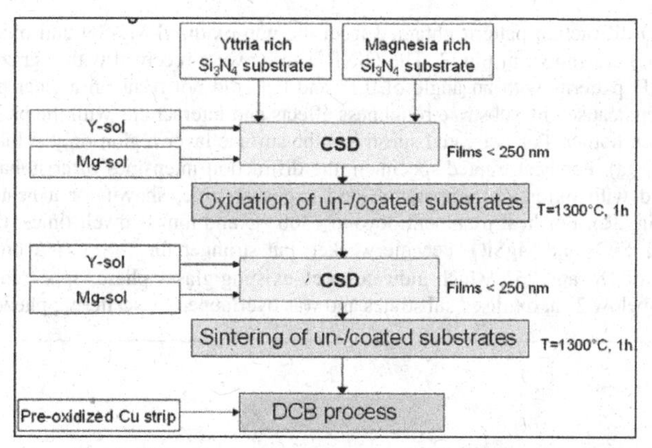

Fig. 1. Flow chart of manufacturing DCB-Si_3N_4 substrates

The DCB samples were subjected to a 90° peel test. The bond strength was defined as average load per unit Cu width [10]. The weight changes (mg/cm^2) of as received, uncoated Si_3N_4 ceramic substrates were measured after the heating cycles as a function of time. The effectiveness of the coating layer to reduce the amount of oxidation of the substrates was investigated by comparing the weight gain of coated and uncoated samples. The cross section as well as the plan view of Si_3N_4 substrates was examined by means of SEM of the type LEO 1530 VP Gemini (Zeiss, Oberkochen, Germany) to define microstructure and coating thickness. A X'Pert pro diffractometer (PANalytical, Almelo, Netherlands) was used for grazing incidence (GI) XRD with an angle of 0.1°, 0.3°, 0.5°, 0.7°, 1°, 2°, 3°, 4° and 5°, respectively, to identify the layer phases at different X-ray penetration depths. Hot stage microscopy (HT16, Hesse Instruments, Osterode, Germany) of the DCB process under nitrogen atmosphere and surface roughness measurements by means of a conventional profilometer (Mitutoyo, Tokyo, Japan) were performed.

RESULTS

Characteristics of the coating sols

Table II summarizes the measured characteristics of the used Y- and Mg-sols for dip-coating.

Table II. Characteristics of coating sols

Composition	Alkoxide	Catalyst	pH	Viscosity (Pa/s) at 100 s^{-1}	Sol (wt. %) Y, Mg
Y3	$Y(OC_4H_9)_3$	Glacial acetic acid	2.1±0.2	0,007±0.002	3
Mg5	$Mg(OC_2H_5)_4$	Glacial acetic acid	4.2±0.2	0,007±0.002	5
Mg3	$Mg(OC_2H_5)_4$	Glacial acetic acid	3.8±0.2	0,004±0.002	3

Uncoated Mg-SN and Y-SN substrates

Fig. 2 shows the isothermal oxidation curves of the uncoated Mg-SN and Y-SN substrates at 1400°C. The oxide growth of Mg-SN follows a clear parabolic rate law, the oxidation started at 1200°C. In contrast, the Y-SN parabolic kinetic behaviour is less pronounced. The material exhibits a higher oxidation resistance, the oxidation did not start until 1300°C.

The GIXRD diffraction pattern obtained from the non-oxidised Mg-SN and oxidised Mg-SN specimens (1300 °C) are shown in Fig. 3. The GIXRD scans were received with a grazing incidence angle of 3°. GIXRD patterns with an angle of 0.1° and 0.3° did not result in a clear pattern due to scattering phenomena caused by substrate roughness effects and interactions with the primary beam as well as the specimen holder. For untreated substrates the surface layer region only exhibit diffraction lines of Si_3N_4 (Fig. 3a). For heat treated specimen, the diffraction intensities of cristobalite, Mg- and Y-silicate increased with oxidation temperature and exposure time, shown for a heat treatment at 1300°C for 1 h (Fig. 3b). For heat treatment beyond 1300 °C and longer dwell times, the diffraction intensities of Si_3N_4, SiO_2 and $MgSiO_3$ became weaker, but stronger for $Y_2Si_2O_7$. A broad reflection between 2θ-angles of 18° and 25°, which indicates a co-existing glassy phase, appeared only slightly for GIXRD-angles below 2° at oxidised substrates and was overlapped by scattering phenomena.

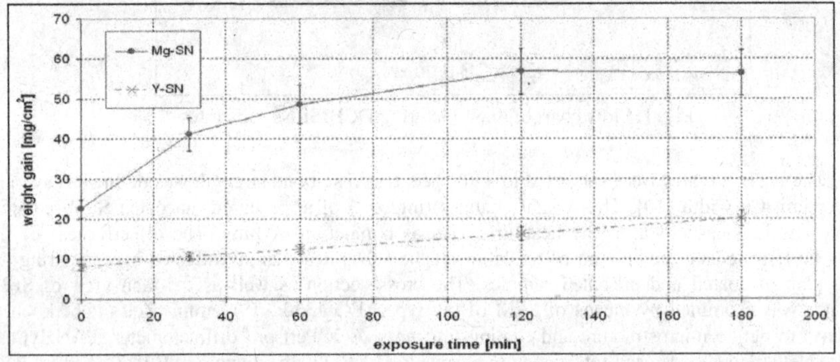

Fig. 2. Isothermal oxidation curves of Mg-SN and Y-SN substrates at 1400°C in air.

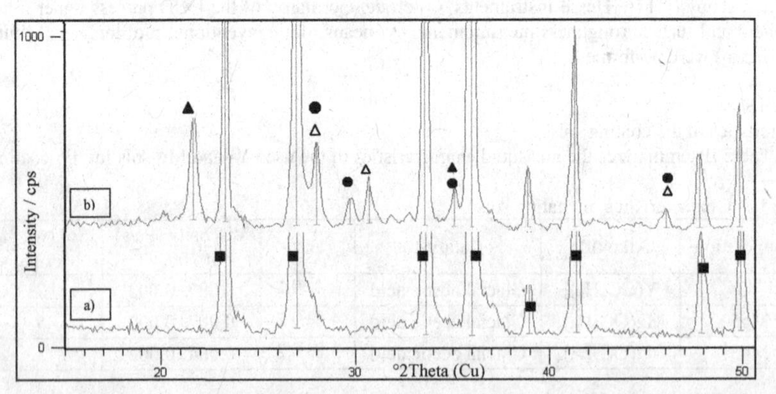

Fig. 3. GIXRD diffraction pattern measured with a GI angle of 3°: a) non-oxidised Mg-SN; b) oxidised Mg-SN, heated at 1300°C for 1h. ■: β-Si_3N_4. Δ: $MgSiO_3$. ●: α-$Y_2Si_2O_7$. ▲: SiO_2 (cristobalite).

Both, uncoated, non-oxidized Y-SN or Mg-SN substrates reveal strong blistering and poor bonding strength of Cu on the substrate after the DCB process.

This is illustrated exemplarily in Fig. 4a for Mg-SN substrates. In case of uncoated Si_3N_4 substrates of both suppliers, which were oxidised twice at 1300°C for 1 h, Cu adhesion was accomplished after the DCB process. The corresponding surface after peel testing is shown in Fig. 4b and 4c, respectively. The bonding strength of Cu on oxidised Y-SN was quite poor compared to oxidised Mg-SN. The cross section of this copper-bonded Y-SN substrate with its thin, porous oxide layer is shown in Fig. 5. The overall thickness (oxide layer + coating layer), coating thickness and the surface roughness were determined (Tab. III, IV).

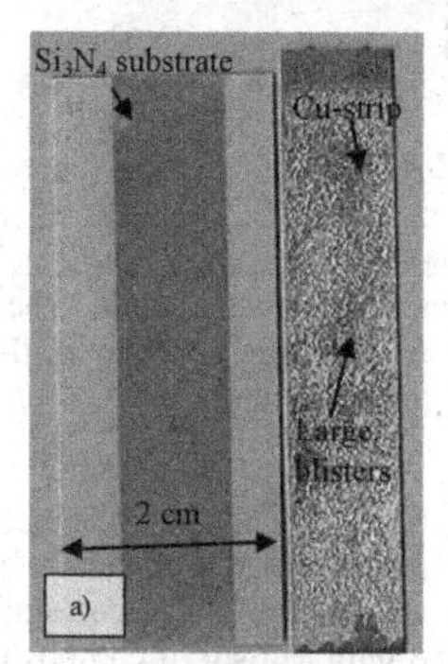

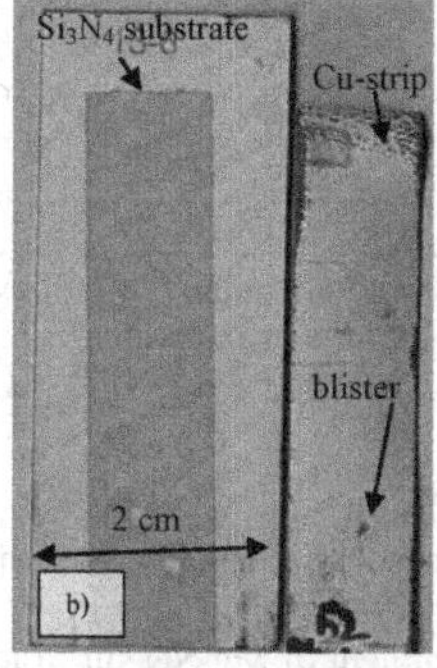

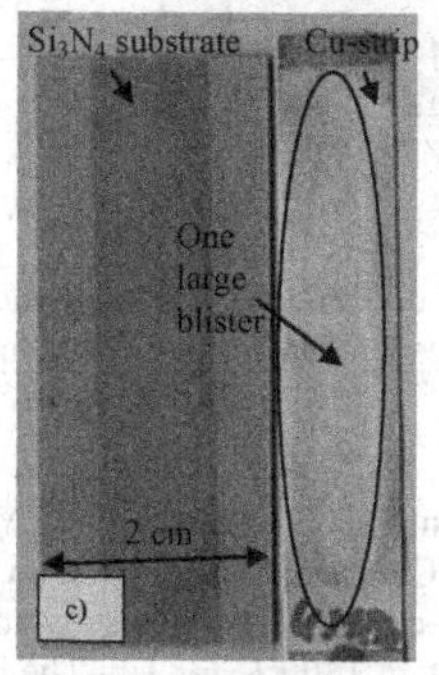

Fig. 4. Uncoated Si_3N_4 substrates + Cu-strip after bonding and peel testing: a) unoxidised Mg-SN; b) oxidised Mg-SN, heated at 1300°C for 1 h (2x); c) oxidised Y-SN, heated at 1300°C for 1 h (2x).

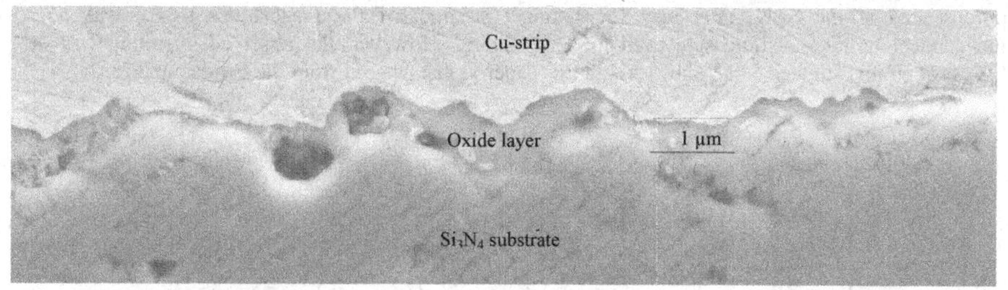

Fig. 5. SEM micrograph of a copper-bonded, oxidised Y-SN substrate, heated at 1300°C for 1h (2x)

Yttrium coated Mg-SN substrates

Dip-coating of MgO rich Si_3N_4 substrates with an Y-sol followed by oxidation at 1300°C for 1 h and re-dip-coating and again sintering at 1300°C for 1 h, resulted in partly dense layers of 1.7 μm in thickness which showed a 60 % increase in oxidation compared to uncoated substrates (Fig. 6a). The illustrated GIXRD scan was taken under a GI angle of 3° which reveals γ-$Y_2Si_2O_7$ beside Si_3N_4, SiO_2, $MgSiO_3$ and α-$Y_2Si_2O_7$ (Fig. 6b). Similar to uncoated oxidised substrates, GIXRD patterns could not be exactly measured at GI angles below 0.7° and the broad peak caused by co-existing glassy phase was relatively weak, too. The bonding strength of Cu, bonded to the top and bottom sides of the substrates was fairly good and identical for both sides, but less good than the strength of Cu bonded to the top side of Mg-SN. Tab. III and IV summarize the peel strengths, blistering behaviour, surface roughness and oxide layer characteristics of the investigated substrates.

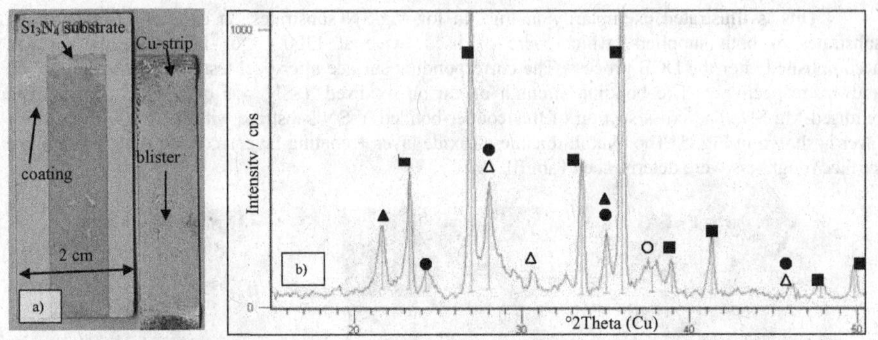

Fig. 6. Left: Mg-SN substrate coated with yttrium sol after bonding and peel testing. Right: The corresponding GIXRD patterns taken under a grazing incidence angle of 3° before bonding. ■: β-Si_3N_4. Δ: $MgSiO_3$. ●: α-$Y_2Si_2O_7$. ○: γ-$Y_2Si_2O_7$. ▲: SiO_2 (cristobalite).

Magnesium coated Mg-SN and Y-SN substrates

Compared with uncoated substrates and independent of the Si_3N_4 substrate manufacturer, both Mg dip-coated substrates showed an enhanced weight gain up to 100% and more by oxidation during sintering at 1300°C for 1 h. The heat treatment of Mg-SN substrate coated with Mg-sol of composition Mg5 led to a thick and opaque oxide layer, as illustrated in Fig. 7a. The characterization of the cross-section yielded Mg-silicate and cristobalite as constituents of the layer and at the top of the coating α-$Y_2Si_2O_7$ crystals (Fig. 7b). The relatively dense layer with a thickness of 2.5 μm shows a narrow porous area at the Si_3N_4 interface. The bonding strength of these substrates to Cu was poor, independent of the substrate side used for Cu bonding. However, the observed blistering was the lowest of all investigated Si_3N_4 substrates in this paper as can be seen from the Cu strip in Fig. 7a.

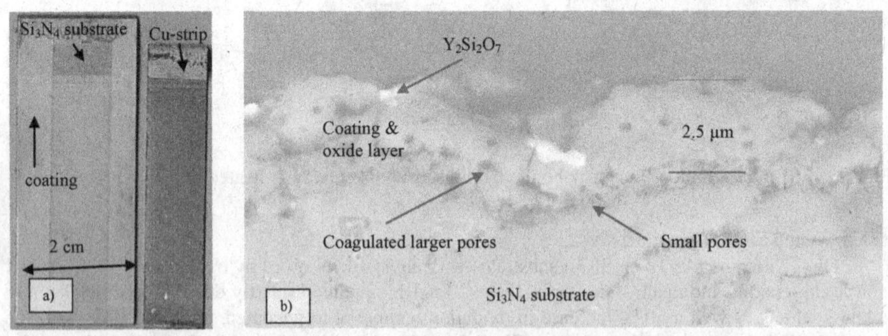

Fig. 7. Left: Mg-SN substrate coated with magnesium sol of composition Mg5 after bonding and peel testing and the corresponding Cu strip. Right: Corresponding SEM micrograph.

Fig. 8a shows the Y-SN substrate coated with a magnesium sol of composition Mg5 and the corresponding Cu-strip after bonding and peel testing. Compared to uncoated Y-SN substrates the peel strength was strongly increased and blistering was clearly impeded. The SEM micrograph of the Cu-bonded substrate (Fig. 8b) exhibits an oxide layer of approximately 1.5 μm thickness with a conglomeration of α-$Y_2Si_2O_7$ crystals in close contact with the Cu-interface.

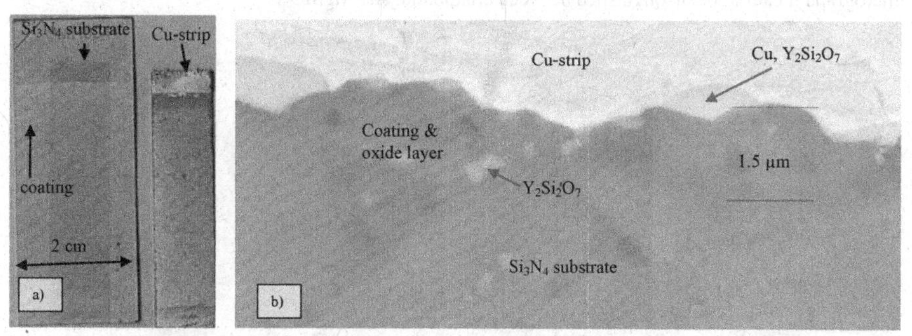

Fig. 8. Left: Y-SN substrate coated with magnesium sol of composition Mg5 after bonding and peel testing. Right: Corresponding SEM micrograph.

Fig. 9a depicts the Mg-SN substrate coated with a magnesium sol of composition Mg3 and the corresponding bonded Cu-strip after peel testing. The bonding strength was nearly similar for Cu stripes bonded to the top or bottom side of these substrates. The strength data were almost as high as the strength of copper, which was bonded onto the top side of non-coated Mg-SN. The copper strip did not indicate any apparent blistering, but tests a in hot stage microscope under DCB conditions with a very thin, full-oxidised copper foil, which was in contact with a coated Mg-SN substrate, revealed the development of some little blisters in the CuO/Cu_2O eutectic melt (Fig. 9b).

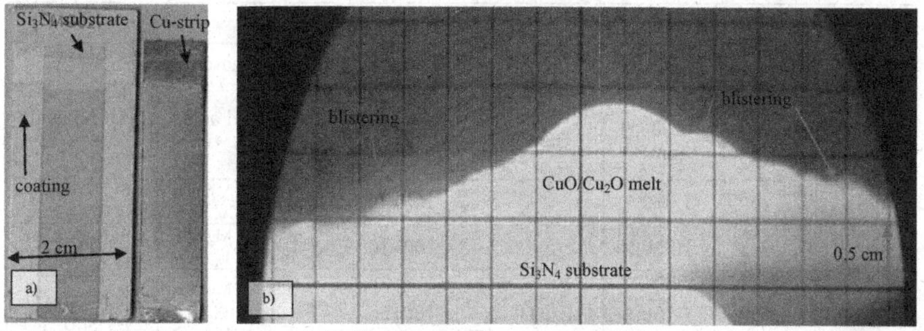

Fig. 9. Left: Mg-SN substrate coated with a Mg-sol of composition Mg3 and Cu-strip after bonding & peel testing. Right: Hot stage micrograph of the Mg-SN substrate in contact with the CuO/Cu_2O-melt.

Microscopically small blisters on the surface of the Cu strip can also be seen in Fig. 10b, which was the counterpart of the interface of the Mg-SN substrate after peel testing (Fig. 10a). Fig. 10a shows that at the interface between the coated Mg-SN substrate and the peeled-off copper exhibits many fine, bright Cu_2O precipitates basically at grain-boundaries close to yttrium, as can be seen from Fig. 8b, too.

In contrast, the plan view of the substrate area exhibits only a few, bright $Y_2Si_2O_7$ crystals. Cu_2O precipitates were also supposed to be around magnesium but unfortunately in the SEM micrograph it cannot be distinguished between cristobalite and $MgSiO_3$.

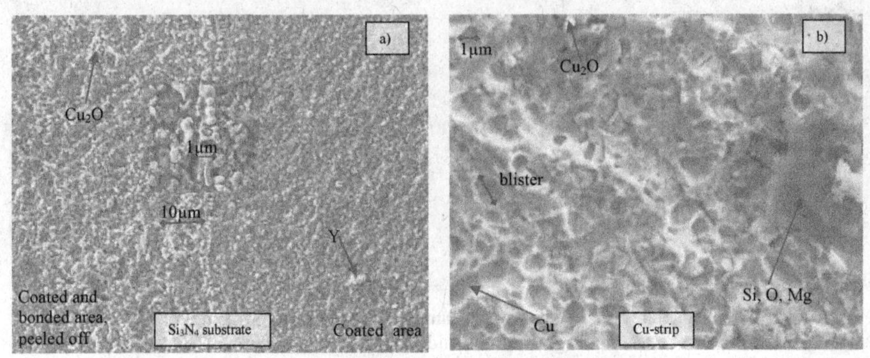

Fig. 10. SEM micrograph of the interface between Cu / Mg-SN substrate, coated with a Mg-sol of composition Mg3, after the peel test. Left: Surface of Mg-SN substrate. Right: Surface of the Cu-strip

Table IV summarizes the data of the peel strength, blistering behaviour and surface roughness for each Mg dip-coated Si_3N_4 substrate. Independent of magnesium or yttrium dip-coating of the Si_3N_4 substrates, the surface roughness was slightly increased by oxidation and sintering at 1300°C for 1 h. Table III gives the corresponding characteristics of the oxide layers of Mg dip-coated Si_3N_4 substrates.

Table III. Characteristics of oxide layers of uncoated & dip-coated substrates, 2x fired at 1300°C, 1h

Si_3N_4 substrate	Coating sol	Oxide layer characteristics				
		Overall layer thickness (µm)*	Coating thickness (µm)	Oxidation ratio (wt%) $\left(=\frac{m_{coated\ substrate}}{m_{uncoated\ substrate}}\right)$	Phase composition**	
					0.3°<GI angle < 0.5°***	0.5°<GI angle < 5°
Y-SN	/	1.0±0.5	/	100±10	β-Si_3N_4, $MgSiO_3$ α-$Y_2Si_2O_7$, SiO_2	β-Si_3N_4, $MgSiO_3$ α-$Y_2Si_2O_7$, SiO_2
Mg-SN	/	1.5±0.5	/	100±10	β-Si_3N_4, $MgSiO_3$ α-$Y_2Si_2O_7$, SiO_2	β-Si_3N_4, $MgSiO_3$ α-$Y_2Si_2O_7$, SiO_2
Mg-SN	Y3	1.7±0.5	0.10±0.02	160±20	β-Si_3N_4, $MgSiO_3$ α-,γ-$Y_2Si_2O_7$, SiO_2	β-Si_3N_4, $MgSiO_3$ α,γ-$Y_2Si_2O_7$,SiO_2
Mg-SN	Mg5	2.5±0.5	0.25±0.05	215±15	$MgSiO_3$ α-$Y_2Si_2O_7$, SiO_2	β-Si_3N_4, $MgSiO_3$ α-$Y_2Si_2O_7$, SiO_2
Y-SN	Mg5	2.0±0.5	0.25±0.05	190±15	$MgSiO_3$ α-$Y_2Si_2O_7$, SiO_2	β-Si_3N_4, $MgSiO_3$ α-$Y_2Si_2O_7$, SiO_2
Mg-SN	Mg3	2.0±0.5	0.15±0.05	190±15	β-Si_3N_4, $MgSiO_3$ α-$Y_2Si_2O_7$, SiO_2	β-Si_3N_4, $MgSiO_3$ α-$Y_2Si_2O_7$, SiO_2

*: Before bonding
**: SiO_2 is always present as cristobalite
***: Due to scattering phenomena a co-existing SiO_2 glassy phase can not be excluded.

Table IV. Bonding strength, surface roughness and blistering behaviour of uncoated and dip-coated substrates, 2x fired at 1300°C for 1h

Si_3N_4 substrate	Coating sol	Bonding strength of top and bottom side of substrate (N/cm)		Blistering behaviour	Surface roughness R_a (μm)
		Top side	Bottom side		
Y-SN	/	20±5	15±5	Strong	0.72±0.05
Mg-SN	/	40±5	20±5	Quite strong	0.56±0.04
Mg-SN	Y3	30±5	30±5	Quite strong	0.73±0.04
Mg-SN	Mg5	25±5	25±5	Moderate	0.76±0.04
Y-SN	Mg5	35±5	30±5	Quite strong	0.74±0.05
Mg-SN	Mg3	35±5	30±5	Moderate	0.75±0.04

DISCUSSION

Uncoated Mg-SN and Y-SN substrates

The kinetic of the weight increase indicated by the isothermal oxidation curves of Fig. 2, especially for non-coated Mg-SN substrates, are controlled by outward diffusion of sintering additives and inward diffusion of oxygen as described by Nickel [15]. The results of the oxidation tests underline the beneficial effect of yttrium addition to Si_3N_4 as also reported by Lange et al. and Wu et al. for hot-pressed Si_3N_4 materials [16,17]. As expected, the weight gain of the investigated Si_3N_4 substrates of high thermal conductivity was significantly larger, due to the higher relationship of surface area to mass but also because of the coarser microstructure and different amounts of sintering additives. In contrast to the different oxidation curves, the GIXRD diffractograms obtained from oxidised, non-coated Mg-SN and Y-SN specimens did not vary in their phase composition, merely in diffraction intensity. Fig. 3 illustrates the alterations of GIXRD patterns which are influenced by the differences in chemistry, crystallinity and topography caused by oxidation of the non-coated Mg-SN as described by Oliveira et al. [9]. However, for all investigated non-coated and coated substrates in this paper the distinct co-existing SiO_2 glassy layer could not be detected as reported for Al_2O_3-doped Si_3N_4 [7]. Besides the detected crystalline phases, oxidation products such as Mg_2SiO_4, Si_2N_2O, $YMgSi_2O_5N$ cannot be excluded due to possible overlaps of their diffraction lines [4,18].

The reason why especially uncoated, but oxidised substrates exhibited a higher bonding strength at the top side than at the bottom side (Table IV) can only be explained by a dissimilar microstructure of tape cast substrates, which leads to a different oxidation behaviour and oxide layer thickness. Consequently, the manufacturing process of Si_3N_4 tapes (casting direction, top and bottom side, design of stacks etc.) plays a very important role for the achieved properties, which is already known for the thermal diffusivity of Si_3N_4 [19]. Nevertheless, for uncoated, non-oxidized Si_3N_4 substrates of both suppliers it does not make any difference because the reaction of the Cu-Cu_2O eutectic melt with the Si_3N_4 grains is so boisterous (Fig. 4 a) that it does not wet the substrate surface. But this is the prerequisite for direct bonding between Cu and ceramics [20]. In consideration of the discrepancy between the achieved peel strength and the blistering behaviour of uncoated, oxidised Mg-SN and Y-SN (Tab. III, IV) it can be supposed that the overall thickness of the oxide layer is crucial and for oxidised Y-SN apparently still too thin (Fig. 4b, c, Fig. 5).

Yttrium coated Mg-SN substrates

To overcome the problem of a too thin oxide layer as well as the different bonding behaviour of the top and bottom side of oxidised substrates, an yttrium coating was applied. Yttrium was chosen due to its ability to react with Cu to form $Y_2Cu_2O_5$ and with SiO_2 to form Y_2SiO_7 [21]. As described by

Chang et al., a strong bond can only be accomplished by two major bonding mechanisms of mechanical anchoring and of forming an interfacial bonding phase [22] Therefore, the heat treatment of dip-coated Si_3N_4 fulfils two important functions. Firstly, calcination, phase transformation and densification of the dip-coating and secondly, oxidation of the substrate in order to cover all Si_3N_4 grains by a thin passivation layer enabling an adhesion via chemical reaction of the dip-coated layer.

Fig. 6b confirms the occurrence of a chemical reaction between Y_2O_3 and the oxidation product SiO_2. Thus, in addition to the common oxidation product of α-Y_2SiO_7, the polymorph γ-Y_2SiO_7 was synthesized which crystallized from the initial amorphous sol-gel coating as reported by Parmentier et al. [23]. The absence of Y_2O_3 indicates a strong reaction between the coating and the oxide layer of the heat-treated substrate, which finally led to a thicker oxide layer (Tab. III) than in case of uncoated Mg-SN substrates. It seems that the 100 nm thin dip-coating layer interfered the passivation mechanism of the SN surface by reducing the viscosity of the passivating glass phase and therefore, the layer did not act as a diffusion barrier, but rather as a promoter of SiO_2 formation and cation migration which resulted in crystallization of γ-Y_2SiO_7 until the entire Y_2O_3 of the dip-coating layer was used. Lange et al. and Tsarenko et al. declared this phenomenon insufficient passivation by an initially impurity-laden oxide layer on Si_3N_4, which causes an incomplete coverage of all unstable phases. This results in further oxidation, SiO_2 crystallization and other phase transformations, causing internal stresses, cracking, pore evolution and disintegration [16,24].

Keeping in mind the deleterious effect of enhanced oxidation on the quality of the coating layer, it has to be brought to mind that all coated substrates in this work were produced using a double dip-coating and heating cycle. The first heat treatment stands for an impurity-dependent Si_3N_4 oxidation and the second for a densification (crack healing) of the preliminarily formed oxide layer and densification of the second coating layer [25].

Table IV shows that the peel strength of yttrium coated Mg-SN substrates was quite high but the blistering behaviour was inferior compared to Mg-SN substrates coated with a Mg-sol of composition Mg3. A possible explanation for these differences could be seen in the lower overall thickness and in the presence of γ-Y_2SiO_7 instead of α-Y_2SiO_7 (Tab. III).

Magnesium coated Mg-SN and Y-SN substrates

According to the afore mentioned theory of an impurity-dependent Si_3N_4 oxidation, the even enhanced weight gain through oxidation of magnesium coated Si_3N_4 substrates was a result of a slightly higher dip-coating thickness (Tab. III). It is generally known that a complete coverage of a surface is obtained, if the layer thickness is 10 times higher than the R_a value of the surface. If this is not fulfilled, it can be assumed that the coating is disrupted into islands. Considering the as-fired surface roughness of the Si_3N_4 substrates of supplier A and B the dip-coating thickness of 250 nm was still not high enough to cover the complete surface (Tab. I). This is also consistent with a rise in surface roughness (Tab. IV), especially for substrates coated with magnesium sol of composition Mg5. For these samples, X-ray GI angles under 0.7° could not be exactly measured due to scattering effects, and no diffraction patterns of Si_3N_4 could be detected at all (Tab. III).

For Mg-SN substrates coated with magnesium sol Mg5, Fig. 7b exhibits typical sub-surface flaws due to exaggerated oxidation. Lewis also describes a two-tier oxide structure which consisted of a crystallized silicate layer at the top of a glassy layer [26]. For the used Mg-SN substrates, oxidation data showed that the second heat-treatment did not only result in densification but also in re-oxidation triggered by an excess of Mg and a first oxide layer which was not dense enough to act as diffusion barrier. Thus, delamination-like cracks developed and decreased the peel strength (Tab. IV). Nevertheless, the blistering effect was moderate due to the high oxide layer thickness (Tab. III).

Unlike Mg-SN, dip-coating of Y-SN substrates with magnesium sol of composition Mg5 yielded a perfectly thick oxide layer (Fig. 8a, b). This is due to the fact that Y-SN substrates showed a larger oxidation resistance than Mg-SN substrates and therefore it required a higher concentration of ions in the coating to generate the optimal oxide layer thickness. The good bonding strength and moderate blistering effects of Y-SN was not only achieved because of an optimal thickness of the oxide layer but also because of the dense microstructure.

Referring to direct-copper-bonded alumina substrates, Holowczak et al. already showed the positive effect of MgO on the DCB substrate quality as well as Kim et al. for the DCB process on MgO-doped Si_3N_4 [10,8]. Also in this study, un-coated, oxidised Mg-SN substrates exhibited a higher bonding strength than Y-SN substrates (Tab. IV). Nevertheless, the reason for the positive effect of MgO is not clear so far. It can be supposed, like for Y_2O_3+Al_2O_3-doped Si_3N_4, that the bonding mechanism is based on the reaction between the Cu-O eutectic melt and the glassy phase [7]. But then the bonding results for Y-SN substrates should have been better than those of Mg-SN substrates. Consequently, an additional bonding mechanism seems to act which is based on the ability of MgO to form an interfacial bonding phase with copper like $CuMgO_2$ or Cu_3MgO_4 and with SiO_2 like $MgSiO_3$ or Mg_2SiO_4 [21]. This could be the explanation why Mg-SN coated with Mg-sol of composition Mg3 yielded the best Si_3N_4 DCB composites (Fig. 9a).

Fig. 9b) and 10b) show that even for the most favourite microstructure and composition of an oxide layer for the DCB process, blistering could not be avoided. Hot stage microscopy under DCB conditions proved that the dense oxide layer was attacked by the eutectic copper oxide melt which thereby was enriched by dissolved layer oxides. SiO_2, in particular, exhibits an eutectic temperature with Cu_2O at 1030°C which is below the bonding temperature of ~1073°C. Therefore, SiO_2 is believed to enrich the eutectic copper oxide melt during bonding [21]. Thus, the bonding mechanism was definitely provided on the one hand by the so-called glass bond theory [27] causing adherence via the direct reaction or solution between the glass components and metal oxides. This is combined with an anchoring effect as a result of the penetration of glass phase into the cavities of Cu and Si_3N_4 substrate (Fig. 10 a, b). On the other hand, the bonding mechanism was probably provided by the strong adherence of the re-crystallized fine Cu_2O residues (maximal 3 µm) on the Si_3N_4 substrate surface comparable with the up to 100 µm long Cu_2O dendrites observed at direct-copper-bonded alumina substrates [22]. The notional thin interfacial film of Cu_3MgO_4 and $Y_2Cu_2O_5$ between the Cu_2O residues and the Si_3N_4 substrate surface was not further investigated in this paper.

CONCLUSIONS

Direct copper bondable coatings were formed on Si_3N_4 substrates of high heat conductivity from two different suppliers by dip-coating with Y- and Mg-based sols followed by thermal treatment. The difference in the amount of sinter additives which migrated into the coating during oxidation and which reduced the viscosity of the remaining glassy phase causes the observed differences in crystallisation behaviour, porosity and thickness of the oxide layers. As a result, different bonding strengths of Cu on these oxide layers were observed. Due to a reproducible formation of an oxide layer on the surface of the Y-SN and Mg-SN substrates by chemical solution deposition of thin sol-gel films a controlled enhanced oxidation could be accomplished. For the manufacture of DCB-Y-SN and Mg-SN substrates, a good bonding and moderate blistering was achieved using an optimized Mg-sol and a double heating cycle. MgO seems to be more beneficial than Y_2O_3 due to Mg-ions which strongly reduce the viscosity of the glass phase and further the formation of thermodynamically stable $MgSiO_3$ crystals. Thus, free SiO_2 is effectively bonded which, for its part, is unstable against the CuO/Cu_2O eutectic melt during the DCB process.

The overall advantage of the formation of Mg-controlled oxide layers on Si_3N_4 substrates is the formation of a dense oxide layer with the optimum thickness of about 2 µm at a temperature of 1300°C. This is an advantage especially for uncoated Y-SN substrates, because a temperature much higher than 1300°C is needed to form a comparable thick oxide layer - such higher process temperatures would degrade the high thermal conductivity and the strength of the SN-substrates.

REFERENCES

[1]Maerz, M., Poech, M. H., Schimanek, E., Schletz, A., Mechatronic integration into the hybrid powertrain – the thermal challenge, Intern. Conf. of Automotive Power Electronics, 1-6 (2006).

[2]Schulz-Harder, J., Advantages and new development of direct bonded copper substrates, Microelectronics Reliability, 43, 359-365 (2003).

[3]Xu, W., Ning, X., Zhou, H., Lin, Y., Study on the thermal conductivity and microstructure of silicon nitride used for power electronic substrate, Mater. Sci. & Eng., B99, 475-478 (2003).

[4]Zhou, Y., Zhu, X., Hirao, K., Lences, Z., Sintered reaction-bonded silicon nitride with high thermal conductivity and high strength, Int. J. Appl. Ceram. Tec., 5, 119-126 (2008).

[5]Greenhut, V. A., Chapman, T. R., Engineering high-quality ceramic-metal bonds, Ceram. Trans., 138, 61-100 (2003).

[6]M. Brochu, Pugh, M. D., Drew, R. A! L., Joining silicon nitride ceramic using a composite powder as active brazing alloy, Mater. Sci. & Eng., A374, 1-2, (2004).

[7]Tanaka, S., Nishida, K., Ochiai, T., Surface characteristics of metal bondable silicon nitride ceramics, Proceedings of International Symposium on Ceramic Components for Engine, 249-56 (1983).

[8]Kim, S. T., Kim, C. H., Park, J. Y., Son, Y. B., Kim, K. Y., The direct bonding between copper and MgO-doped Si_3N_4, J. Mater. Sci., 25, 5185-5191 (1990).

[9]Oliveira, M., Agathopoulos, S., Ferreira, J. M. F., The influence of Y_2O_3-containing sintering additives on the oxidation of Si_3N_4-based ceramics and the interfacial interactions with liquid Al-alloys, J. Eur. Ceram. Soc., 25, 19-28 (2005).

[10]Holowczak, J. E., Greenhut, V. A., Shanefield, D. J., Effect of alumina composition on interfacial chemistry and strength of direct bonded copper-alumina, Ceram. Eng. Sci. Proc, 10, 1283-1294 (1989).

[11]Takashi, T., Silicon nitride circuit board and producing method therefore, EP Pat.No. 0798781 (1997)

[12]Honkanen, M., Minnamari, V., Lepisto, T., Oxidation of copper alloys studied by analytical transmission electron microscopy cross-sectional specimens, J. Mater. Res., 23, 1350-1357 (2008).

[13]Choi, H., Hwang, S. Sol-gel-derived magnesium oxide precursor for thin-film fabrication, J. Mater. Res., 15, 842-845 (2000).

[14]Ning, H., Ma, J., Huang, F., Wang, Y., Preoxidation of the Cu layer in direct bonding technology, Appl. Surf. Sci., 211, 250-283 (2003).

[15]Nickel, K. G., Corrosion of non-oxide ceramics, Ceram. Int., 23, 127-133, (1997)

[16]Lange, F. F., Singhal, S. C., Kuznicki, R. C., Phase relations and stability studies in the siliconnitride (Si3N4)-silicon dioxide-yttrium oxide pseudoternary system, J. Am. Ceram. Soc., 60, 249-52 (1977)

[17]Wu, C. CM., McKinney, K. R., Rice, R. W., Mcdonough, W. J., Freiman, S. W., Oxidation weight gain and strength degradation of silicon nitride with various additives, J. Mater. Sci., 16, 3099-104 (1981)

[18]Tanaka, S., Itatani, K., Hintzen, H. T., Delsing, A. C., Oka, A., Effect of silicon nitride addition on the thermal and mechanical properties of magnesium silicon nitride ceramics, J. Eur. Ceram. Soc., 24, 2163-2168 (2004)

[19]Li, Bincheng, Pottier, L., Roger, J. P., Fournier, D., Watari, K., Hirao, K., Measuring the anisotropic thermal diffusivity of silicon nitride grains by thermoreflectance microscopy, J. Eur. Ceram. Soc., 19, 1631-1639 (1999)

[20]Wittmer, M., Stimmell, J., Strathman, M., Materials Issues in Silicon Integrated Circuit Processing, Mater. Res. Soc. Symp. Proc., 40, 441-447, (1984).

[21]Roth, R. S., Phase diagrams for ceramists, Cumulative index for volumes I-V, 1969-1984.

[22]Chang, W., Greenhut, V. A., Shanefield, D. J., Johnson, L. A., Gas-metal eutectic bonded Cu to Al_2O_3 substrate-mechanism and substrate additives effect study, Ceram. Eng. Sci. Proc., 14, 802-812 (1993).

[23]Parmentier, J., Bodart, P. R., Audoin, L., Massouras, G., Thompson, D. P., Harris, R. K., Goursat, P., Besson, J.-L., Phase transformations in gel-derived and mixed-powder-derived yttrium disilicate, $Y_2Si_2O_7$, by X-ray diffraction and Si MAS NMR, J. Solid. State. Chem., 149, 16-20 (2000).

[24]Tsarenko, I. V., Du, H., Lee, W. Y., Effects of additive and impurity species on the oxide morphology of silicon nitride, J. Mater. Res., 18, 878-884 (2003).

[25]Choi, S. R., Tikare, V., Pawlik, R., Crack healing in silicon nitride due to oxidation, Ceram. Eng. Sci. Proc. 12, 2190-2202 (1991).

[26]Lewis, M. H., Barnard, P., Oxidation mechanisms in Si-Al-O-N ceramics, J. Mater. Sci., 15, 443-448 (1980).

[27]Floyd, R. J., Effect of composition and crystal size of alumina ceramics on metal-to ceramic bond strength, Am. Ceram. Soc. Bull., 42, 65-70, (1963)

APPLICATION OF SEMICONDUCTOR CERAMIC GLAZES TO HIGH-VOLTAGE CERAMIC INSULATORS

André L. G. Prette and Vincenzo M. Sglavo
DIMTI, Università degli Studi di Trento
Trento, TN, Italy

Orestes E. Alarcon and Marcio C. Fredel
CERMAT, Universidade Federal de Santa Catarina
Florianópolis, SC, Brazil

ABSTRACT

The flashovers that occur in electrical insulators are problems that can shut down the transmission and distribution electrical energy line. Insulators placed in regions with high salinity or atmospheric pollution are more susceptible to failure by occurrences of flashover. In the present work the effect of the application of a semiconductor coating containing 95 wt% SnO_2 and 5 wt% Sb_2O_3 to high-voltage electrical insulators was studied. The coating was applied by dip coating. The effects of semiconductor coating are ohmic heating and homogeneous electrical current distribution on the surface of the insulator. The properties of developed glazes were analyzed by SEM, XRD, surface resistivity and salt spray chamber tests. The thermal expansion coefficient of the coating was evaluated and matches the bulk one.

The flexural strength of the insulator was increased by approximately 13% due to the lower thermal expansion coefficient of the semiconductor glaze which creates surface compressive stresses. Tests in salt-spray chamber showed that semiconductor glazes can develop surface heating of about 55°C and endure large quantities of deposited salt, under high-voltage, without the development of flashovers, thus increasing the performance of the insulators.

INTRODUCTION

Due to the passage of electric current through the air around a high-voltage insulator a flashover can be formed. The escape current produced by the deposition of impurities and salt suspension in the air over the insulators can cause the shutdown of the distribution energy network in industrial or coastal regions. Because the high wettability of the ceramic insulator's surface, this impurities can form a layer of pollution and humidity, an electrolyte, which in certain conditions can originate a high escape current, or flashover, and activate the self protection system of the energy supply, shutting it down[1,2,3].

Figure 1 shows a schema of the flashover formation in a general electric insulator. Figure 2 shows how the semi-conductor glaze acts avoiding the flashover phenomena.

There are two potential approaches to avoid the problem of flashover: (i) avoid the presence of contamination or (ii) avoid the presence of humidity.

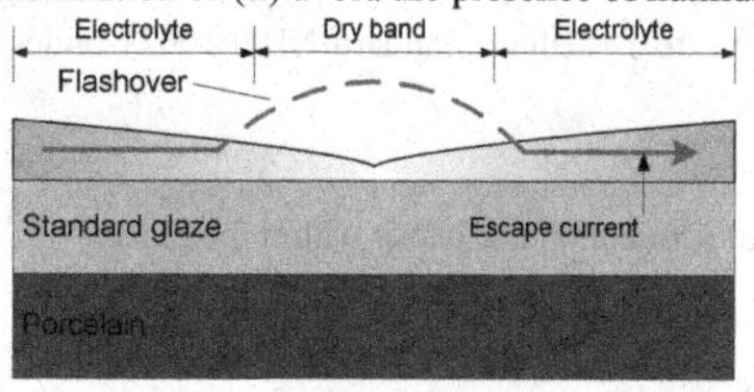

Figure 1. Escape current schema in a general high-voltage electric insulator.

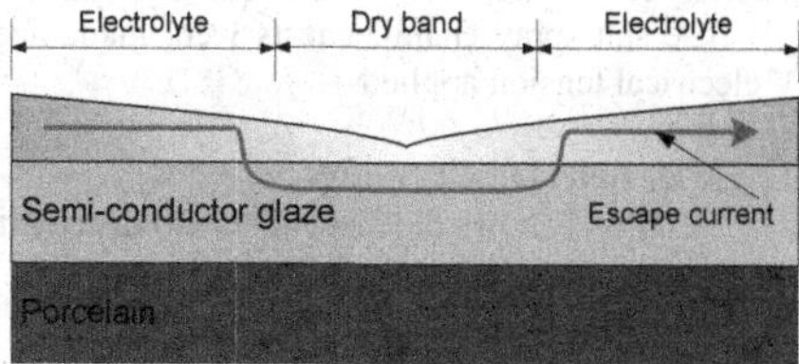

Figure 2. Escape current schema in a semi-conductor high-voltage electric insulator.

Nowadays the solution applied is to avoid the presence of contamination by the periodic maintenance of the insulator, washing them[2,3]. This situation demands long operation time, high costs and it is not always efficient because even with periodic washing, flashovers are often observed in the transmission line[1,2,4].

The aim of the present work is to modify the insulator surface to increase its performance by adding semiconductor compounds in its glaze formulation. The used compounds are antimony oxide (Sb_2O_3) doped tin oxide (SnO_2) that were added to the standard transparent glaze industrially used. The idea is to promote an ohmic heating and a better electric current distribution[5,6].

Among the characteristics needed for an efficient semiconductor glaze one can include the adequate resistivity to allow a Joule heating and better uniformization of tension over the material[7]. The values suggested for surface resistivity must vary around 11-200 [MΩ/□][5,8,9].

MATERIALS AND METHODS

A standard glaze (ETG, Germer) typically applied in industry was used as reference sample to compare surface quality and physical properties (mechanical resistance, thermal expansion coefficient, surface resistivity) of the developed glazes. The semiconductor powder was prepared mixing 95% of SnO_2 (purity 99,96%) with 5% of Sb_2O_3 (purity 99,96%), such proportion results in a highest electrical conductivity for this material. Raw materials were milled and homogenized, calcined at 1050°C for 3 h, according to previous works[4,5,10,11].

The semiconductor glaze was obtained by mixing the standard glaze with 30 wt% of the semicondutor powder. The glaze was applied on porcelain substrates (IEC 672 C-120, Germer) by dip coating with immersion for 5 s in the slurry and successive drying and firing at 1250°C for 7.5 h.

Scanning electron microscopy analyses were made in a SEM Philips DL-30 with energy dispersive x-ray (EDXS) analyzer. The semiconductor properties of the glaze was qualitatively analyzed during SEM observation: the samples were partially coated with a conductor material; in this way, if the samples were semiconductor they could carry the small current generated by SEM electron beam and avoid the phenomena called charging-up. Figure 3 shows the coating procedure: normally insulating samples must be full coated and grounded to avoid the electrons accumulating over their surface and generate the charging-up phenomena. Therefore, only half of the surface was coated the other remaining uncoated.

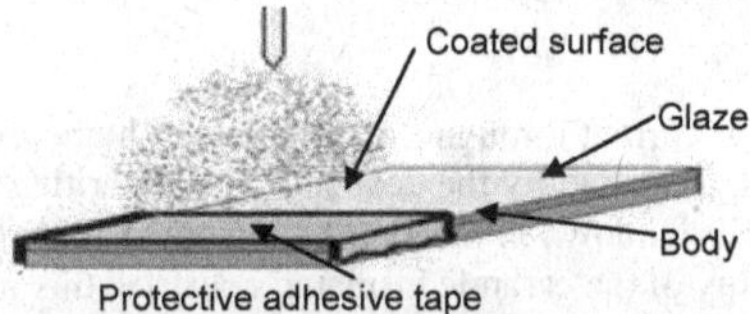

Figure 3. Coating schema realized for SEM analysis[12].

Dilatometric analysis were carried out with heating rate of 10°C/min independently on the body, the standard glaze and the semiconductor glaze. Cylindrical samples with 20 mm length and 5 mm diameter were used. The mechanical resistance tests were performed by 3-point flexural test. Surface resistivity tests were carried out by parallel bars [13].

The salt-spray chamber tests were made according to Brazilian Standard NBR 10621 under 13 kV electrical tension applied.

RESULTS AND DISCUSSION

Figure 4 shows the dispersion and distribution of semiconductor phase within the glaze.

Figure 4. Micrograph of the surface of the semiconductor glaze.

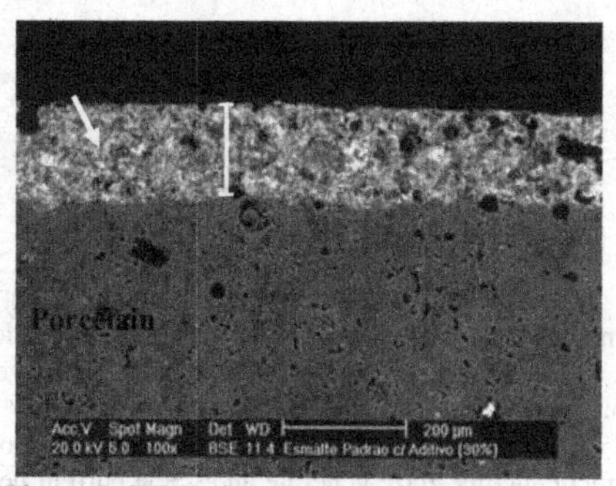

Figure 5. SEM micrograph of transversal section of the semiconductor glaze.

In Figure 5 it is possible to observe that the thickness of the coating is about 150 μm; it is homogeneous and the semi-conductor phase (lightest color) is homogenously dispersed over the glaze thickness.

SEM analysis allowed to make a preliminary measurement of semiconducting property of the glaze by the coating technique showed in Fig. 3. Figure 6 and Figure 7 show two distinct behaviors related to charging-up phenomena. In Figure 6 (standard glaze) it is not possible to analyze the uncoated surface, because it is charging-up. Conversely, in the semiconductor glaze (Figure 7) the surface can be observed in detail since no charging-up phenomena are present and the semiconductor glaze can carry the small current (80 mA) generated by SEM.

Figure 6. SEM micrograph showing the standard glaze in a preliminary conducting test.

Figure 7. SEM micrograph showing the semiconductor glaze in a preliminary test.

Table I shows that the standard glaze has smaller thermal expansion coefficient than ceramic body and the semiconductor glaze presents an even smaller thermal expansion coefficient. This difference is favorable for creating compressive stresses in the glaze and increasing the mechanical strength.

Table I. Thermal expansion coefficient of body and studied glazes.

	α value(t – 325 °C)
Ceramic Body	51.2×10^{-7} °C^{-1}
Standard Glaze	50.8×10^{-7} °C^{-1}
Semiconductor Glaze	49.2×10^{-7} °C^{-1}

Table II. Mechanical strength of body and glazes (values in [MPa]).

	Average	Std. Dev.
Insulator Porcelain	66,6	4,2
Standard Insulator	77,6	5,7
Semicond. Insulator	87,8	9,1

From Table II it is possible to observe that the smaller thermal expansion coefficient of the semiconductor glaze accounts for a greater mechanical strength.

The surface resistivity of semiconductor glaze measured by parallel bars [13] is equal to 70 MΩ/□ that characterizes the material as a semiconductor.

The salt spray chamber tests were carried out under extreme conditions. The insulators with semiconducting glaze generate an escape current nearly constant about 3.5 mA and the insulators with standard glaze showed 0.2 mA and failed after 110 and 135 hours of test.

The semiconductors insulators also develop the smallest third harmonic current indicating low flashover activity. A surface temperature of 55°C due to the Joule effect was also measured. As shown in Figure 8 and Figure 9 a consistent salt deposit was formed over the surface of semiconductor insulator although they did not failed during the test period.

Figure 8. Insulator with semiconductor glaze after 72 hours of salt spray chamber test.

Figure 9. Insulator with semiconductor glaze after the end of salt spray chamber test.

CONCLUSIONS

Homogenous antimony oxide (Sb_2O_3) doped tin oxide (SnO_2) dispersion was obtained in industrially transparent glaze which showed evident seimconductor character. The lower linear thermal expansion coefficient of the semiconductor glaze led to a higher mechanical strength of the insulator.

Even with an abnormal salt accumulation, the insulators with semiconducting glaze were fully operative, which indicates a longer lifetime when applied in normal conditions. No flashover was detected in the insulators with semiconducting glaze. Production of high-performance electrical insulators with semi-conductive properties is possible and viable with no change in the existing industry production lines.

REFERENCES

[1]D. M Leite and A. G .Kanashiro, Projeto de Pesquisa em isoladores poluídos. Revista Mundo Elétrico, São Paulo, 1985.
[2]M. R. Lima, Personal communication. 2006.
[3]J. Liebermann: Refractories and Industrial Ceramics, v.43, n.1-2, p.55-64. 2002.
[4]E. O. Abdelaziz, *et al*: Control, Communications and Signal Processing, 2004. First International Symposium on , vol., no., pp. 493-497, 2004
[5]C. Fontanesi, et al.: Journal of the Europan Ceramic Society, 1998. 18: p. 1593-1598.
[6]R. Aguiar, et al.: Cerâmica, 2004. 50: p. 134-137.
[7]S. Matsui, et al.: in Proceedings of the 5th International Conference on Properties and Applications of Dieletric Materials. 1997. Seoul, Korea.
[8]G. Rosenblatt: *Semiconducting Glaze Composition*, U.S. Patent, Editor. 1976.
[9]N. Higuchi, K. Takayuki, and S. Nagoya, *Electrical Insulators*, U.S. Patent, Editor. 1978.
[10]A. Ovenston: Journal of Materials Science, 1994. 29: p. 4946-4952.
[11]H. Ullrich and S. Gubanski: IEEE Transactions on Dielectrics and Electrical Insulation, 2005. 12(1): p. 24-33.
[12]A. L. G. Prette, Desenvolvimento de esmaltes cerâmicos: aplicação em isoladores elétricos de alto desempenho, Florianópolis, 2006.
[13]W. Maryniak, et al.: Trek Application Note, 2003. p. 1-4.

CERAMICS FOR ABRADABLE SHROUD SEAL APPLICATIONS

Dieter Sporer
Sulzer Metco
Hattersheim, Germany

Scott Wilson
Sulzer Metco
Wohlen, Switzerland

Mitchell Dorfman
Sulzer Metco
Westbury, NY, USA

ABSTRACT

Significant improvements in the overall efficiency of thermal turbomachines can be achieved by reducing leakage flows in compressor and turbine modules. One approach to minimize leakage losses is the use of thermally sprayed coatings that function as so called abradable seals. In the hottest section of turbomachines, such seal coatings need to be manufactured from ceramics to be able to withstand the high temperature environment of modern engine designs. This paper reviews the design, manufacture and performance of ceramic abradable seals with varied chemistry, structure and properties of the seal coating. Emphasis is put on seal designs for operating temperatures up to 1200 °C (2190 °F) and cost effective solutions.

INTRODUCTION

Just a 1 percent increase of efficiency for the 2500 GW installed electricity base worldwide leads to a reduction of CO_2 emissions of 300 million tonnes per year, with savings of 100 million tonnes of fossil fuel[1]. For gas turbine engines, a 1 percent efficiency increase can be achieved by properly managing the tip clearances in the high pressure turbine section[2]. Passive clearance control using abradable seals can provide such clearance management. The high temperatures encountered in the turbine stages require ceramic materials to provide for durable hot gas path sealing. With thermal spraying, a relatively simple and cost effective method of providing ceramic seal coatings is available. For maximum service temperatures up to 1200 °C (2190 °F), seals can be manufactured from zirconia based ceramics. Abradable zirconia coatings have to meet a number of different requirements: Coatings need to be relatively thick to accommodate a rotor incursion depth. This challenges the cyclic resistance of the structure and coatings need to be designed to show excellent resistance to thermal cycling. To provide for this, several approaches can be used that will be discussed in the following chapters. To fulfill their function as an abradable seal, these coatings need to show the ability to be cut by fast rotating turbine hardware. Several different design approaches to meet the desired abradability and durability of the seals can be employed.

TIP CLEARANCE LEAKAGE LOSSES

Figure 1 shows a hypothetical high pressure turbine stage of a gas turbine engine with unshrouded turbine blades. δ is the clearance between the rotating turbine blades and the stationary shroud of the turbine. Although necessary by design, this clearance can have a disproportionately high influence on the stage efficiency. In general, an increasing tip clearance causes stage performance deterioration. The losses associated with the clearance are called tip leakage losses. As part of the overall blade losses, the turbine tip clearance loss arises because at the blade tip the working gas does not follow the intended flow path and therefore does not contribute to the turbine power output and interacts with the shroud wall boundary layer. Depending on the specific engine design considered, tip clearance losses can account for up to 40 percent of the total losses[3].

The leakage flow is induced by the pressure difference between the pressure and the suction side of the blade leading to a gas flow from the region of higher pressure to that of a lower pressure. This flow interferes with the main flow along the turbine blade from the leading to the trailing edge to form a tip clearance vortex. This tip clearance loss is called indirect tip leakage and it is mainly influenced by the clearance gap size and the difference in pressure on the suction and pressure side and hence the aerodynamic loading of the blade. The direct tip leakage is the through-flow of gas that occurs in the tip clearance from the higher pressure, upstream of the blade, to the lower pressure region, downstream of the blade. The flow loss phenomena are schematically represented in Figure 1.

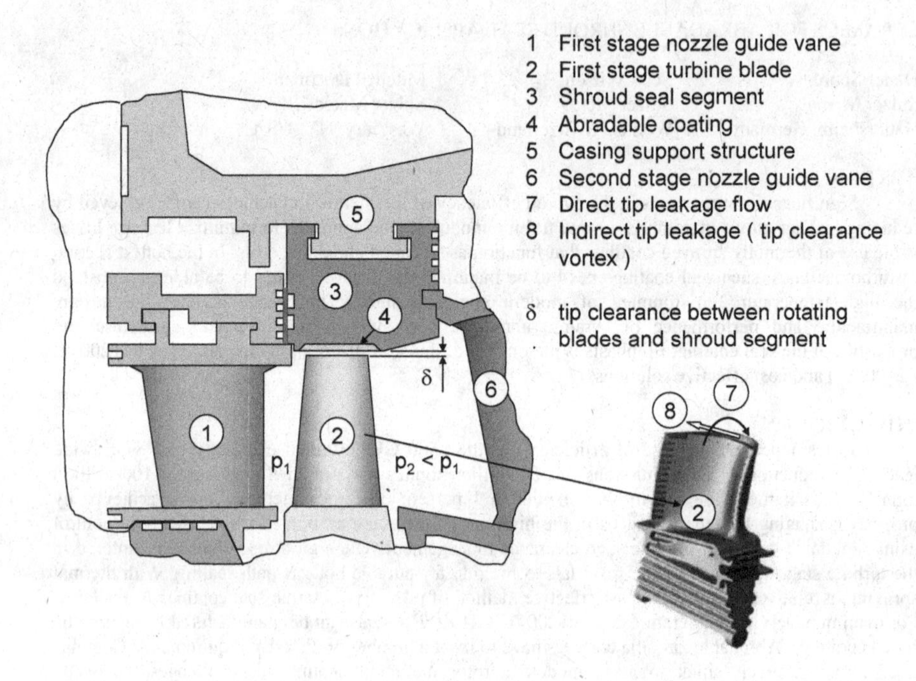

Figure 1. Section through a hypothetical gas turbine engine high pressure stage showing leakage flows over the tip of un-shrouded blades.

The magnitude of tip clearance losses have been quantified by various authors. Typically an increase of the tip gap size equal to 1 percent of the blade span causes around 2 percent drop in stage efficiency for turbine designs with shroudless blades that have a design clearance of the order of 1-2 percent of the blade span[3]. Numerical studies carried out to investigate the effect of tip geometry on the tip leakage flow and heat transfer characteristics in un-shrouded axial flow turbines revealed the following: A clearance of 1 percent of the blade span causes 1-2 percent of the primary flow to leak and hence reduces stage efficiency by 1-3 percent[3]. Three-dimensional computational fluid dynamics (3D CFD) applied in modeling the flow losses of a first stage, high pressure turbine of a stationary gas turbine showed that by reducing the hot operating tip clearance by 90 percent, both the stage efficiency and power output could be improved by 3 percent, leading to an improvement of engine overall efficiency and power output of 1 percent[2]. Slightly better than the predicted improvements were measured during actual operation of a similar gas turbine engine having the reduced hot clearance[4].

The influence of the turbine tip clearance on the efficienc of a gas turbine engine was simulated with a simple tip clearance loss model and the results correlated well with the performance of an existing engine[3]. In this model the efficiency is calculated by

$$\eta = \eta_{ref} [1 - k.(\delta / h).(r_t / r_m)] \quad (1)$$

where $k = 1 + 0.586 (\psi_{ztip}^{3.63})$
η is the rotor efficiency with tip clearance effects included
δ is the tip clearance height
h is the blade span
ψ_{ztip} is the tip Zweifel loading coefficient

Overall the savings from reduced tip leakage losses resulting from reduced tip clearance δ are massive. For a modern industrial gas turbine, a 1 mm (0.040 in) reduction in stage 1 and stage 2 tip clearances can be worth in excess of $ 1 million with the hotter first stage typically contributing two thirds of the total improvements[5].

TIP CLEARANCE REDUCTION USING ABRADABLE COATINGS

A simple method of reducing tip clearances is the use of so called abradable coatings. These can be employed to minimize the operating tip clearance by allowing the blades to cut into them while keeping wear and heat generation in the blade tip to a minimum. Abradable coatings can be applied directly on the stationary turbine shroud components as shown schematically in Figure 1. With thermal spraying a relatively simple and cost effective method of applying the abradable coatings is available. In order to provide an effective seal, the abradable coating must fulfill a number of requirements. In short these can be summarized in the two main requirements abradability and durability, which are often conflicting.

Ideal abradability as one of the major requirements is schematically explained in Figure 2 for a hypothetical turbine or compressor stage. Upon a rotor displacement as shown, which may be caused by main shaft bending, the rotor comes into contact and overlaps with the casing. In the case of the ideal abradable casing lining (case I), the rotor can cut into the stator without any reduction in rotor diameter or blade length. Upon returning into the initial position, the design tip clearance δ_0 is maintained around the circumference with the exception of a small sickle shaped area that now shows increased tip clearance. Case I is a boundary case characterized by all seal and no blade wear. In the case of a non-abradable casing (case II), the rotor blades are worn and reduced in length which gives rise to an increased tip clearance $\delta_{1,II}$ which is much larger in size than the initial design tip clearance δ_0 along the entire circumference. This results in stage efficiency deterioration as explained above. Case II is also a boundary case characterized by all blade and no seal wear. In most practical cases, mixed forms with some seal and some blade wear will be observed. In general, blade wear up to 5 percent of the total rotor / stator geometrical overlap, the so called total incursion depth, will be acceptable.

Rub interactions can have many different causes. A rotor eccentricity as a result of rotor shaft bending as shown in the example in Figure 2 can be a result of manoeuvre loads, heavy turbulences or hard landing. Rub interactions also occur during engine transient conditions such as start-up, acceleration, deceleration and hot re-start. To get an idea of the likely overlap or total incursion depth sizes, it is possible to make an approximation as to how much a turbine disc, carrying the turbine blades, will expand when heated to say 600 °C (1100 °F). For a 600 mm (23.6 in) disc made of a Ni alloy it turns out that the thermal expansion would be up to 5 mm (0.20 in). In addition to this there is an elastic stretch from the centrifugal loading of around 1 mm. The casing against which the rotating components seal will obviously also see some thermal growth, but it will not be at the same time because of the thermal masses involved, and of course, it will not see the centrifugal growth at all. In some cases the ingenuity of design engineers can be brought to work with thermal management to continuously counteract all of these variables and provide a satisfactory "active clearance control". In most cases, however, the potential risk for a contact is acknowledged and an abradable seal fitted for "passive tip clearance control".

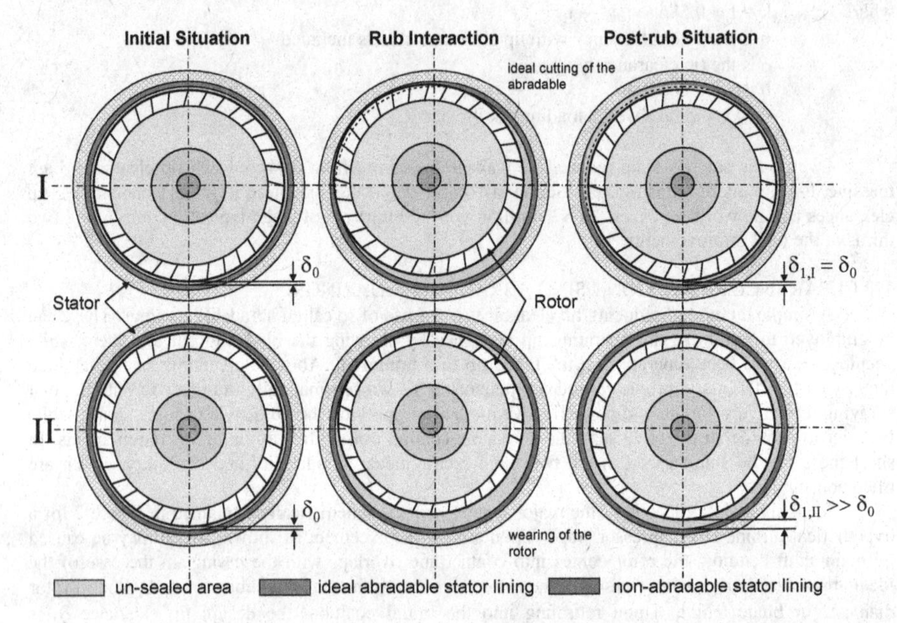

Figure 2. Evolution of tip clearance for a rub situation caused by rotor displacement for the boundary cases of all seal wear (case I) and all rotor wear (case II).

Clearance control applications in the hottest section of gas turbine engines, the first turbine stages, requires the use of materials that can withstand the challenging environment. Therefore ceramic abradable materials will have to be used to cope with the extreme temperatures seen in modern engines. While ceramics offer a distinct advantage over metallic materials in terms of their hot gas oxidation resistance, they challenge the design of seal coatings from other points of view including their cuttability. How ceramics can be cut by metal blades is partly explained by the basic wear or cutting mechanism at high relative speeds as explained in the following.

The mechanisms of wearing or cutting of a ceramic abradable at high speed in engine applications is accomplished by ejection of small particle debris, typically < 100 μm in size, behind the moving blade[6]. This partly sets the criteria for the design of such materials. The ideal cutting mechanism of ceramic abradables as dicussed here is shown in Figure 3: As a rotating blade passes parallel to the coating material it also incurs perpendicularly towards the coating, due to thermal expansion and/or mechanical deformation of shafts or casings. Eventually, the blade will impact onto the surface of the abradable coating. This results in a momentum transfer from the blade to surface protrusions of the coating (a perfectly smooth coating surface is not desirable). Initially this means that the surface particles of the coating are pushed away from the incurring blade tip and into the coating. The elastic energy, which is created in this interaction, then pushes the particle towards the surface of the coating, the blade now having passed. If the elastic energy is sufficient to overcome the bond strength between the particle and its neighboring particles, it will be released as debris from the coating, taking with it most of the initial impact energy. It can be seen from Figure 3 that Particle I is being forced into the coating material as momentum is transferred to it from the rotating and incurring blade tip. Particle II has been impacted by the blade tip slightly earlier and is still traveling into the coating but is no longer in contact with the blade. Particle III has stopped moving into the coating and has accumulated enough elastic energy to start moving back towards the surface of the coating. Particle IV is accelerating away from the coating, due to the stored elastic energy of the abradable, and is about to de-bond from the coating. Particles V and VI have already de-bonded and have been released by the coating as debris.

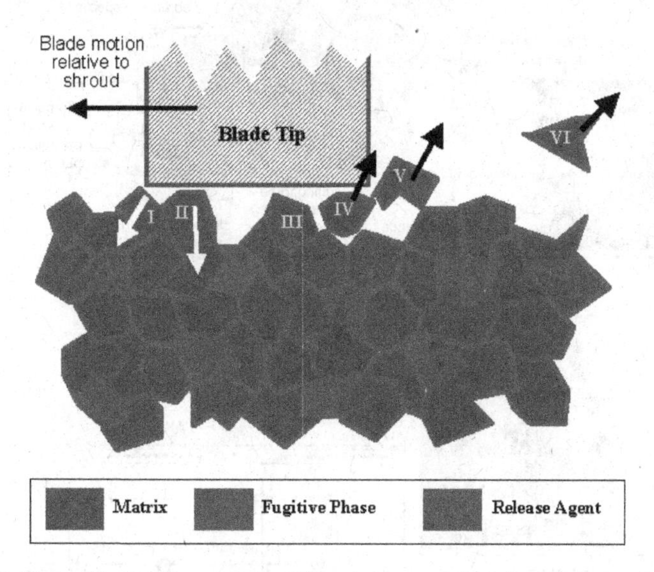

Figure 3. Ideal cutting/wearing of an abradable coating consisting of a ceramic matrix (blue), a fugitive filler or porosity (green) and a network of a release agent phase (red).

TESTING CERAMIC COATINGS FOR ABRADABILITY

Before abradable coatings can be used in engines, they undergo a number of functionality tests including testing for their abradability. The challenge with ceramic abradables lies in the fact that abradability testing needs to take place at relevant application temperatures. In particular abradability test rigs capable of providing testing at up to 1200 °C (2190 °F) need to be selected. Such a rig is installed at Sulzer Innotec and shown in Figure 4. This rig uses a high velocity gas stream to heat the abradable specimen to the required temperature. Test blades are mounted on a turbine disc that can produce blade tip speeds of up to 500 m/s (1640 ft/s). A stepper motor allows the controlled moving of the abradable test specimen towards and into the bladed rotor. Incursion rates, the speed at which the specimen is moved into the rotor, can be adjusted from 1 to 2000 µm/s (4×10^{-5} to 2×10^{-2} in/s). Real turbine blades or blade dummies as shown in Figure 5 can be tested. While the use of original blades provides the advantage of closely simulating a specific stage design, the use of dummies with well defined geometry is the preferred set-up for systematic investigations as discussed here. The tip thickness of the dummies can be varied, however, a fixed, rectangular tip design of 2 mm (0.079 in) thickness was used for all tests discussed in this paper.

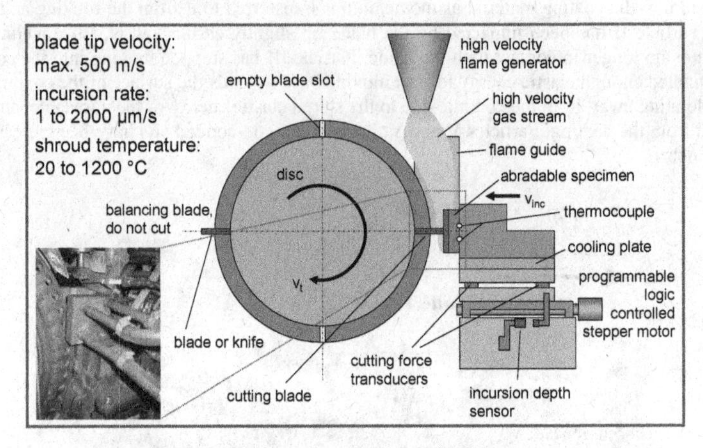

Figure 4. High temperature abradability test rig.

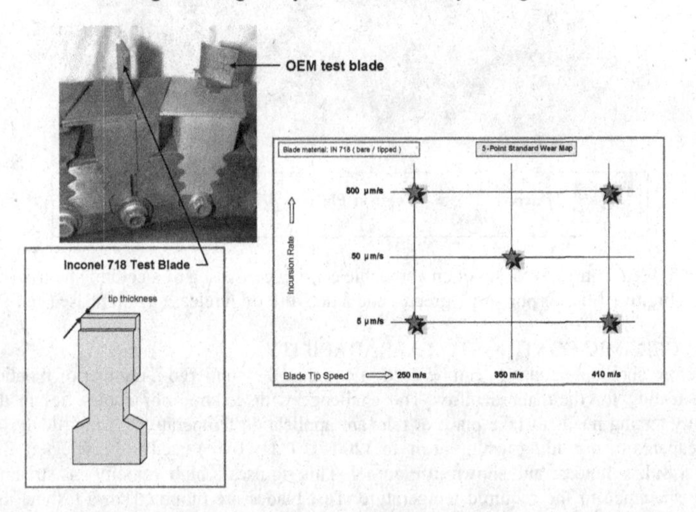

Figure 5. Dummy and OEM test blades. Standard wear map with five test conditions.

The main result derived from the abradability tests is the blade wear which is reported as a percentage of the total incursion depth, which was kept at 0.7 mm (0.028 in) for all tests. While the expected blade tip speeds for rub interactions in engines can be estimated with some certainty, the corresponding incursion rates are mostly not known. Therefore a general screening test makes use of a standard wear map consisting of five different tip speed / incursion rate pairings as shown in Figure 5. Blade wear results need to be interpreted along with a macroscopic assessment of the generated rub path. While high blade wear is normally always indicative of sub-optimal abradability, low blade wear does not necessarily mean optimal abradability performance as the seal coating may show macro-rupture as shown in Figure 6. Such coating rupture may cause large and deep pockets of the seal to be removed early in the rub process and leave no or little material behind for subsequent blade passes to cut into, thereby feigning small blade wear. Another typical but unfavorable wear mechanism observed for ceramic abradables is material transfer from the blade to the ceramic seal. The transferred metal forms an oxidized and hard deposit on the ceramic surface and typically prevents clean cutting of the abradable as a result of the metal tip now rubbing against a dense metal layer. Figure 6 gives an overview of various undesirable coating abradability mechanisms in high-speed rub tests at temperatures between 800 and 1100 °C (1470 and 2010 °F).

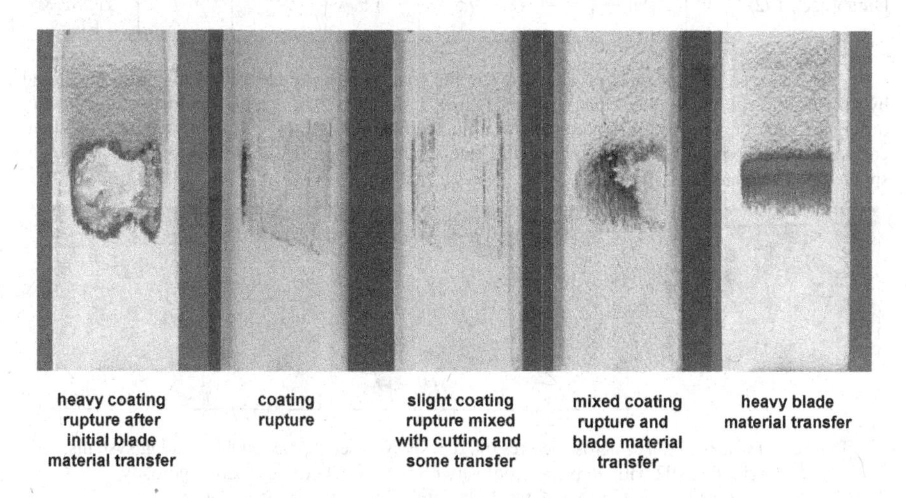

Figure 6. Coating rupture and blade material transfer during high-speed, high-temperature rubbing of zirconia-based seal coatings.

DESIGN OF CERAMIC ABRADABLE COATINGS

General Coating Structure and Materials

As for many metal based abradable coatings, the coating porosity is the driver for the performance of ceramic seal coatings. However, in contrast to metal based abradables, porosity in ceramic based abradables will not only improve abradability but also determines the thermal-cyclic performance of the coatings, as discussed in the following chapters. The materials use partially stabilized zirconia, for example yttria partially stabilized zirconia (YPSZ), as a matrix former. A fugitive phase that will initially be deposited in the as-sprayed coating structure, but can be burned off by a subsequent heat treatment, is often added to help create more coating porosity. Polymer or polymer-based compounds are used as fugitive filler phases. Table I shows the compositions and method of manufacture of commercial ceramic abradable materials. Powders are provided by blending a plasma densified (HOSP) ceramic phase with a compound of polymer and hexagonal boron nitride

(Sulzer Metco 2395 and Durabrade 2192) or by spray drying zirconia, stabilizer and polymer into one composite powder particle (Sulzer Metco 2460 NS). These powders are sprayed by atmospheric plasma spraying (APS) to produce an abradable coating. The seal coating system typically consists of an oxidation resistant metallic bond coat, such as Amdry 962 (Ni 22Cr 10Al 1Y) or Amdry 995C (Co 32Ni 21Cr 8Al 0.5Y), over which the ceramic top coat is deposited.

Table I. Nominal chemical composition of commercial thermal spray powders for ceramic seals.

Material	ZrO_2 (wt%)	Y_2O_3 (wt%)	Dy_2O_3 (wt%)	Polymer (wt%)	Binder (wt%)	hBN (wt%)	Method of Manufacture
Sulzer Metco 2460 NS	Bal.	7.5	-	4.0	4.0	-	Spray Dry
Sulzer Metco 2395	Bal.	7.5	-	4.5	-	0.7	HOSP & Blend
Durabrade 2192	Bal.	-	9.5	4.5	-	0.7	HOSP & Blend

For a given powder material composition, the coating porosity can relatively easily be altered by varying the spray processing parameters[7]. Figure 7 shows typical coating microstructures having different levels of porosity. A feature of these microstructures is a high amount of sinter-resistant, coarse-clustered porosity over and above the fine inter-splat porosity that is typical for thermally sprayed zirconia materials.

Figure 7. Typical coating microstructures of APS-sprayed ceramic abradable coating systems. Durabrade 2192 with 46 percent porosity (left) and 30 percent coating porosity (center) and Sulzer Metco 2395 with 24 percent porosity (right).

Influence of Spray Processing Parameters on Coating Properties

With varied coating porosity, other coating characteristics like hardness, erosion resistance and thermal shock life will change as well, as shown in Figure 8. By varying the material type and the particle size of the ceramic phase in Sulzer Metco 2395 (SM 2395 MC contains a coarser size YPSZ phase, SM 2395 MF a finer one) in combination with applying different levels of plasma power, adjusted by the arc current, the coating porosity of the ceramic top coats can be varied in a wide range. In general the higher plasma power spray settings correspond with lower top coat porosities, higher hardness (HR15Y according to ASTM E18), higher erosion resistance (ASTM G76) and lower thermal shock life in furnace cycle tests at 1150 °C (2100 °F) peak temperature[4].

Optimized Thermal Shock Resistance By Varying Coating Material Chemistry

To accommodate the rub incursion depth, ceramic abradable coatings need to be sprayed relatively thick with a typical top coat thicknesses in excess of 1 mm (0.04 in) as shown in Figure 7. This has an effect on their thermal-cyclic life, as typically the cyclic life is drastically reduced with increasing coating thickness[8]. By using alternative stabilizer phases, the thermal shock properties of coatings can be improved significantly. Through replacing the yttria stabilizer, as used in SM 2395, by dysprosium oxide (Dy_2O_3) in combination with a low-impurity zirconia matrix, the coating thermal shock resistance can be improved by a factor of up to four in the thermal cycle test as described above[4,8]. The benefit is the highest for coating porosities in the > 25 percent regime. This concept is realized in the form of the Durabrade 2192 abradable coating material.

With the different material formulations and a simple, single parameter variation of the spray process, a tool package is provided that allows the design of coatings with broadly varied properties. This can be used to tailor the coatings to the needs of specific turbine stage seal requirements.

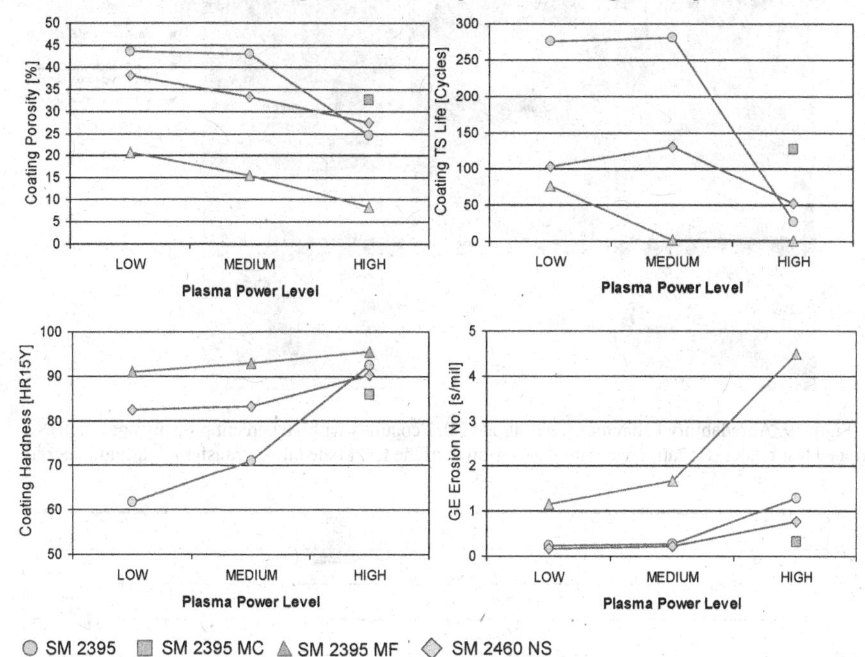

Figure 8. Influence of spray processing parameters on coating characteristics. Coatings have a thickness of 1 mm (0.040 in) and are produced by APS using a Sulzer Metco 9MB plasma gun.

INFLUENCE OF COATING POROSITY ON ABRADABILITY PERFORMANCE

Hard Tipped Blades

As the ceramic coating can be abrasive to bare metal blades, a blade tip treatment is often applied to improve their cutting behaviour. Blade tips and knife edges can be tipped with relatively dense oxide ceramics (Al_2O_3 or ZrO_2) or can have a tipping with hard particles. A commonly used blade tipping employs relatively coarse cubic boron nitride (cBN) particles in an oxidation resistant MCrAlY matrix. To provide for optimal cutting behaviour, the particles stand out from the metal matrix holding them in place. Typically, cBN tipping provides for excellent cutting behaviour as shown in Figure 9. However, if the ceramic abradable is too dense, the particles tend to be pulled out with a corresponding loss of cutting efficiency giving rise to massive blade wear as shown in Figure 10.

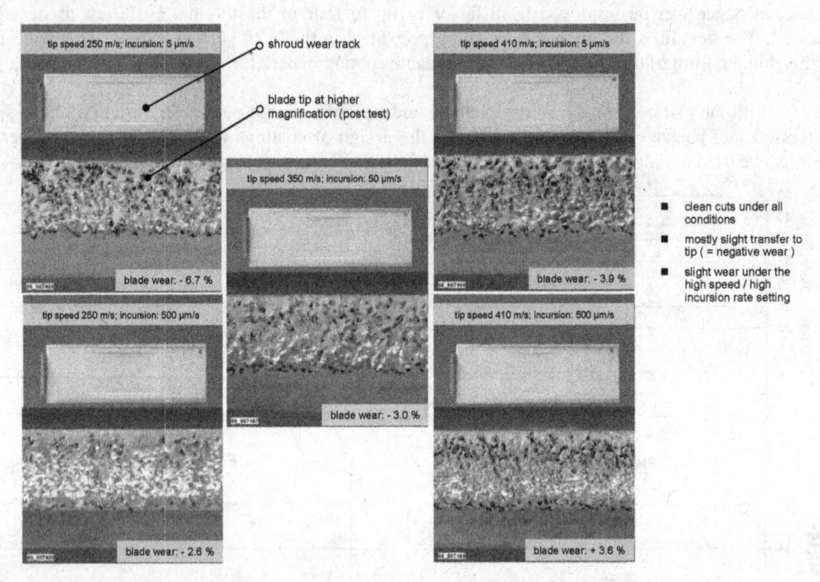

Figure 9. Abradability behavior of Durabrade 2192 coatings with 30 percent porosity against cBN tipped test blades. Negative wear indicates growth of the test blade due to transfer of shroud material.

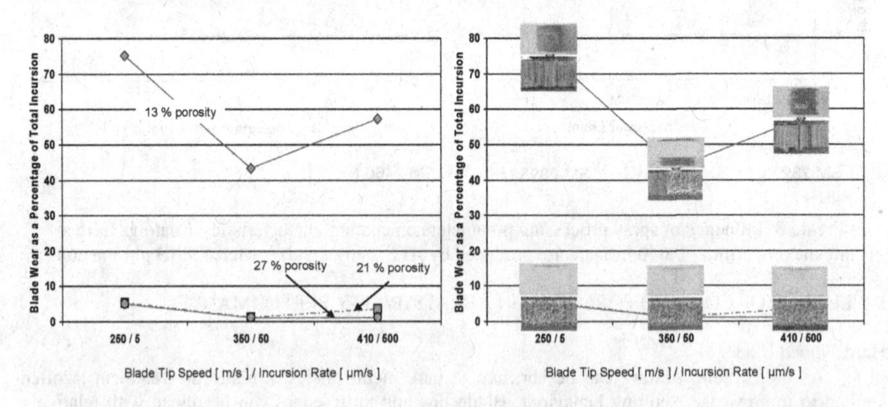

Figure 10. Abradability behavior of SM 2395 type coatings with varied porosity. cBN-tipped blades.

While for certain regimes of coating porosity, cBN–tipped blades cut ceramic abradables very nicely, this approach has two disadvantages: Firstly, the cBN does not have long term oxidation stability at temperatures exceeding 900 °C (1650 °F) and therefore is only effective for cutting during initial start-up rubs and not after long-term, high temperature service. Secondly, the tipping with cBN particles is costly. Therefore the use of un-treated, bare metal blade tips is preferred, provided other specific performance requirements allow for this.

Un-treated, Bare Metal Tips

A number of abradability tests have been performed to assess under what conditions bare, un-tipped blades can rub against zirconia-based ceramic abradables while still meeting a general 5 percent maximum blade wear criterion and providing cutting with little or no coating rupture and/or blade material transfer. As can be expected, the coating porosity is the key enabler for such behavior. Hcanowever, the blade tip speed can play an important role as well, as can be seen in Figure 11.

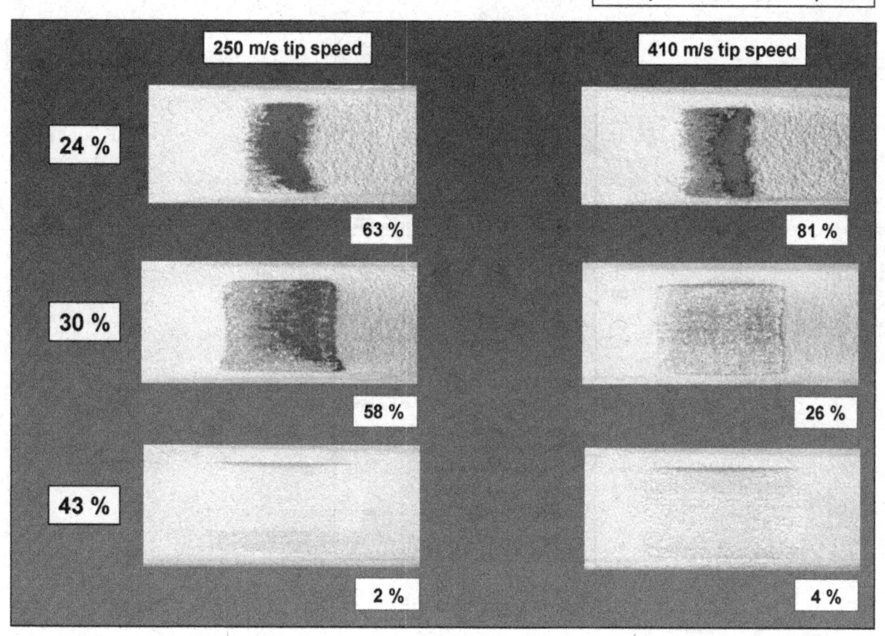

Figure 11. Blade wear as a function of coating porosity and blade tip speed for tests performed at 1100 °C (2010 °F) using bare Inconel 718 blade dummies and a 500 µm/s incursion rate. Top and bottom: Sulzer Metco 2395; Middle: Durabrade 2192.

No significant difference in performance is seen for coatings with a porosity of about 25 percent. Blade wear is very high and accompanied by heavy blade transfer. The short rub path indicates that most of the incursion depth of 0.7 mm (0.028 in) is accommodated by wearing the blade. If the coating porosity is increased to about 30 percent, the coating starts to show relatively clean cutting at the higher blade tip speed of 410 m/s (1345 ft/s), however, the blade wear is still at 26 percent. At the lower speed of 250 m/s (820 ft/s) the blade wear is high and accompanied by some blade material transfer. At a porosity level of just over 40 percent, clean cutting of the abradable with no coating rupture or blade material transfer is observed. This is reflected in the low blade wear values and, overall, indicates almost ideal abradability.

Similar plots can be produced for the lower incursion rate tests. The results are best summarized in a wear map plot as shown in Figure 12. Generally speaking, the abradability improves with reduced incursion rate. However, for the low porosity coating, there is little difference in the blade wear values but a tendency towards heavier blade transfer and less coating rupture with increasing incursion rate. Likewise, little difference in performance is seen for the high porosity coating. For the intermediate porosity, a clear trend towards less blade material transfer, lower blade wear and clean cutting is seen with reduced incursion rate. This condition provides a good compromise of erosion resistance and abradability for a coating systems that will rub at high tip speeds with low or moderate incursion rates.

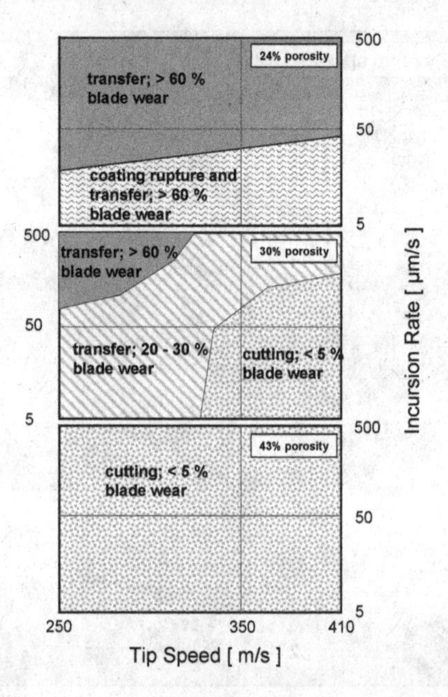

Figure 12. Coating abradabilty as a function of blade tip speed and incursion rate for three different levels of coating porosity.

The abradability performance needs to be balanced with the required durability, given by the erosion and the thermal cyclic resistance, of the seal system. Figure 13 shows an overview chart for the abradability versus erosion requirements for systems using tipped blades and bare metal tips. The chart suggests that the Durabrade 2192 composition provides a better selection for seal systems that are intended for use against un-tipped blades.

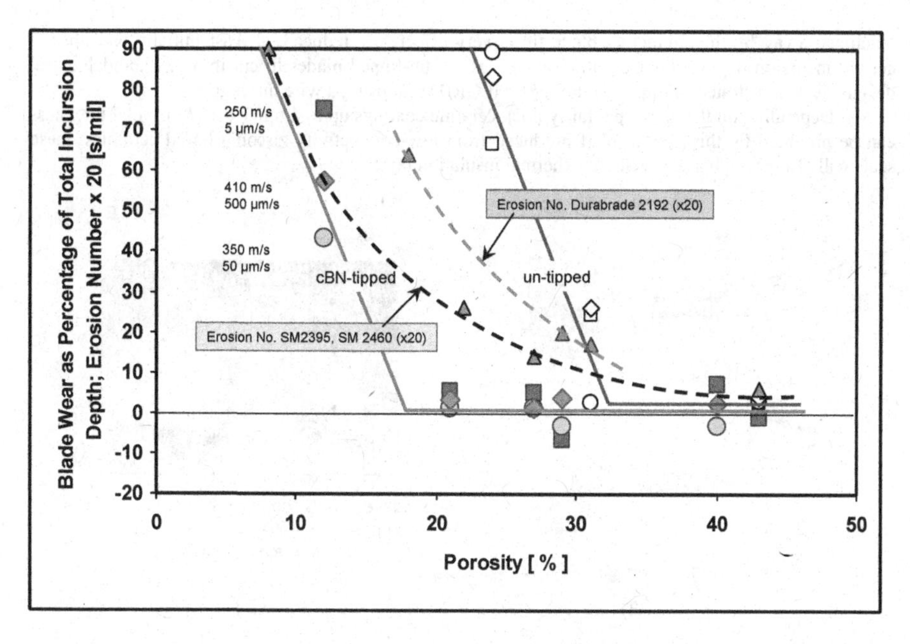

Figure 13. Abradability and erosion performance of various ceramic abradables in dependence on coating porosity.

SUMEGRID – AN ALTERNATIVE DESIGN APPROACH

Instead of depositing the ceramic on a relatively smooth surface, which requires a high degree of porosity in the ceramic to meet the required thermal-cyclic resistance as discussed in the previous chapters, a different design approach can be followed by depositing the ceramic on a structured surface. The structuring of the surface can function as a strain isolator and firmly anchor the ceramic. It can further provide vertical segmentation of the ceramic as can be seen in Figure 14b. Like the columnar structure of EB-PVD coatings, the segmented APS structure can accommodate thermal expansion mismatch, giving it good thermal shock resistance, while at the same time sufficient thickness for rub incursions is provided. In this concept the segmentation takes over the effect of porosity in the previously discussed designs. This allows for the ceramic to be sprayed relatively dense with the advantage that filler-free zirconia materials such as Metco 204NS (ZrO_2 $8Y_2O_3$) can be used. However, any of the standard ceramic abradable materials as listed in Table I can be used for spraying the top layers to create a more abradable outer layer.

Figure 14a shows a selection of surface structures evaluated as strain isolator grids in what is called the SUMEGRID ceramic shroud abradable concept. The structures shown can either be joined to the surface by brazing or be provided as an integrally cast grid as shown in Figure 15. Apart from their functionality of improving the durability of this ceramic abradable system, the grid structure can also serve the purpose of providing a desirable aerodynamic effect, very much like a honeycomb structure in a labyrinth seal configuration. This is based on the fact that the coating will follow the three-dimensional shape of the underlying grid and form isolated cells as seen in Figure 15. This also has a positive influence on the abradability of the structure, as the effective coating area that has to be machined away by the contacting blade tip is very effectively reduced, at least initially and up to a certain incursion depth. This can allow for the use of un-tipped blades to cut into this abradable seal design, as demonstrated in Figure 16 for a SUMEGRID seal using a wire mesh grid.

Depending on the grid type, fairly thick ceramic coatings up to 4 mm (0.157 in) total thickness can be provided by this approach. If produced from low-conductivity zirconia based ceramics, these seals will also provide a very effective thermal insulation layer.

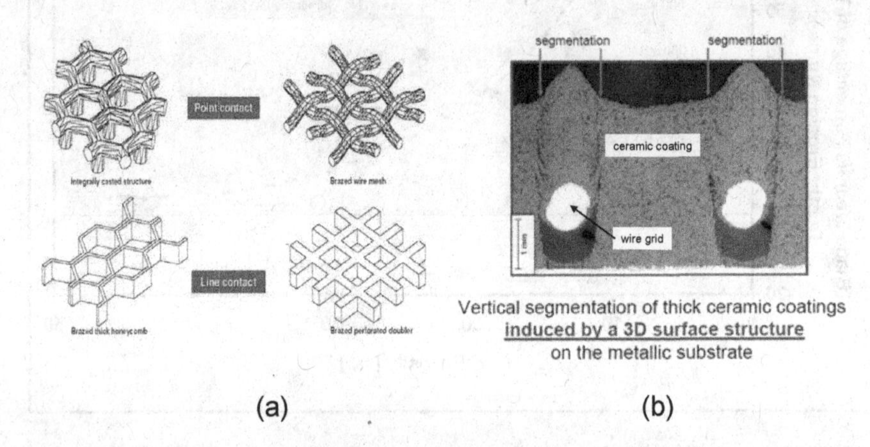

Figure 14. Various types of basic grid structures used in SUMEGRID (a) and vertical segmentation of thick zirconia coatings using a wire mesh metallic grid (b).

Figure 15. Integrally cast hexagonal grid (left) and grid sprayed with zirconia ceramic to provide a 3D-structured abradable surface (right).

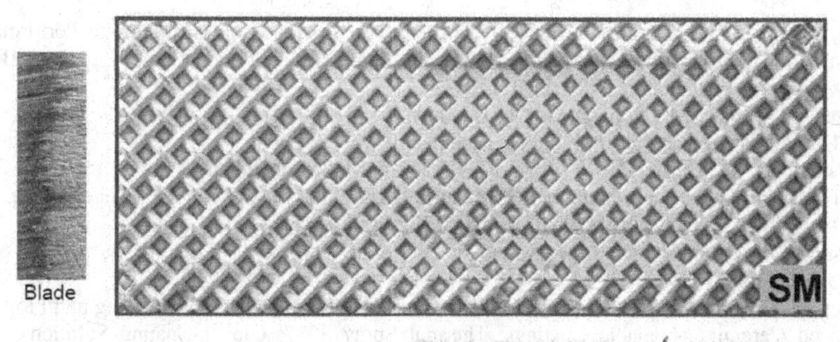

Blade Tip Velocity: 370 m/sec
Blade Thickness: 3.17 mm Incursion Rate: 2 μm/sec
Temperature: 1.000°C Incursion Depth: 1 mm

Figure 16. SUMEGRID zirconia abradable rubbed by an un-tipped Ni-base blade dummy to a total incursion depth of 1mm (0.040 in).

CONCLUSION

Reducing tip clearance leakage losses in turbomachines can significantly improve engine efficiency. Such an improvement can be achieved by minimizing the tip clearance size by applying passive clearance control with abradable coatings. This is most effectively applied in the hottest section of gas turbines, the first two turbine stages. There, just a small reduction in tip clearances can save enormous amounts of fuel and significantly reduce CO_2 emissions. The temperatures encountered in the high pressure turbine section of modern engines requires ceramic materials to provide for durable abradable seals. Various zirconia-based abradable materials and concepts have been developed for this purpose and are now available to engine designers.

Starting from an expected maximum incursion depth for new stage designs or a maximum build-up thickness for repair and up-grade applications, a necessary seal thickness is derived. This will determine the overall seal system design approach, with three-dimensionally structured surfaces typically allowing for larger seal thicknesses as compared to smooth shroud surfaces. Subsequently the required seal durability (erosion and thermal shock resistance) will determine the level of coating porosity for smooth surface designs for an optimal abradable seal life. Finally, the selected seal design needs to be validated in rig abradability testing at the relevant application temperatures before it can be put into service in fleet engines.

A package consisting of materials, basic design approaches, test methods and spray processing parameters to tailor coating properties to specific engine needs is available.

REFERENCES

[1]Information Brochure issued by the European Association of Gas and Steam Turbine Manufacturers hosted by VDMA; EUTurbines (2008).

[2]I. Giovannetti, M. Bigi, M. Giannozzi, D. R. Sporer, F. Capuccini and M. Romanelli, Clearance Reduction and Performance Gain Using Abradable Material in Gas Turbines, Proceedings of ASME Turbo Expo 2008: Power for Land, Sea and Air, June 9-13, 2008; Berlin, Germany, Paper No. GT2008-50290, ASME (2008).

[3]C. Bringhenti and J. R. Barbosa, Effects of Turbine Tip Clearance on Gas Turbine Performance, Proceedings of ASME Turbo Expo 2008: Power for Land, Sea and Air, June 9-13, 2008, Berlin, Germany; Paper No. GT2008-50196, ASME (2008).
[4]D. Sporer, A. Refke, M. Dratwinski, M. Dorfman, I. Giovannetti, M. Giannozzi and M. Bigi, Increased Efficiency of Gas Turbines, Sulzer Technical Review 2/2008.
[5]R. Subramaniam, A. Burns and W. Stamm, Advanced Multi-Functional Coatings for Land-Based Industrial Gas Turbines, Proceedings of ASME Turbo Expo 2008: Power for Land, Sea and Air, June 9-13, 2008, Berlin, Germany; Paper No. GT2008-51532, ASME (2008).
[6]R. Schmid, New High Temperature Abradables for Gas Turbines, Ph.D. Thesis, Swiss Federal Institute of Technology (1997).
[7]D. Sporer, M. Dorfman, L. Xie, A. Refke, I. Giovannetti and M. Giannozzi, Processing and Properties of Advanced Ceramic Abradable Coatings, Thermal Spray 2007: Global Coating Solutions, ASM International, Materials Park, Ohio, USA (2007).
[8]D. Sporer, I. Giovannetti, M. Dorfman, M. Bigi and M. Giannozzi, Structure Property Relationships in Ceramic Abradable Coatings, Proceedings of the Turbine Forum 2008 – Advanced Coatings for High Temperatures, April 23 – 25, 2008, Nice – Port St. Laurent, France, Forum of Technology (2008).

WEAR RESISTANCE OF HARD MATERIALS IN DRILLING APPLICATIONS

Jing Xu1, Hendrik John, Andreas Krafczyk
Baker Hughes, Celle 29221, Germany

ABSTRACT

The world-wide use of hard materials in the oilfield industry encounters the challenges with increasingly complex and costly drilling operation. Focus has been made into engineering ceramics and tungsten carbide cermet as drilling tool hardfacing materials, due to the continued advancement of drilling technology and high demand on wear resistance performances. The wear behaviors of seven types of hard materials have been systematically investigated at two different wear conditions, namely the wet sand/rubber wheel abrasive wear and the high speed particle/water jet erosive wear. Ranking of wear resistance among those hard materials is made for industrial specification and evaluation purpose. Results show that WC-3Co exhibits the highest wear resistance and Al_2O_3 is of the lowest wear resistance among all of the materials evaluated. Abrasive wear testing results are in accordance with erosive wear testing results, regarding to the wear resistance ranking. Results showed that size ratio between hard materials grain size and erodent/abrasive particle size plays role in wear characteristics by affecting the wearing mechanisms. For the tungsten carbide cermet system, WC-Co, the influence of binder content on the hardness and wearing behaviors is discussed. Results show that the higher was the binder content, the lower were the hardness and wear resistance. The variance in wear resistance coincides with a change in the mechanism of materials removal. The present study indicates that the wearing mechanism is mainly plastic deformation, bulk gouging, cutting and removal. However, fragmentation and microchipping of material grains also occur with impact of erodent/abrasive particle size. The research achievements will allow one to establish the procedures for the evaluation and selection of hard materials in drilling applications and help to optimize the current development strategy of engineering ceramics and tungsten carbide cermet.

1. INTRODUCTION

Research on hard materials, engineering ceramics and tungsten carbide cermet, has made significant progress in the last 15 years, driven by the requirements from a variety of industrial applications. [1, 2] For oil and gas drilling industry, due to the continued advancement of drilling technology and high demand on wear resistance performances, to improve the abrasion resistance and erosion resistance of the applied hard materials shows higher priority than other mechanical and thermal properties in drilling applications under specific working conditions. High attention has been made when the wear resistance is studied, especially when taking a variety of well fluid abrasive types into account.

This paper evaluates the response of five types of engineering ceramics and two types of tungsten carbide-cobalt cermet subjected to two substantially different wear test-systems, namely the wet sand/rubber wheel abrasive wear and the high speed particle/water jet erosive wear. The objective is to elucidate the materials type/microstructure/property trends governing the response of these hard materials to each type of wear situation. The base materials are considered in particular detail and the effects of the two tribosystems are compared and contrasted. Ranking of wear resistance among those hard materials is made for the purpose of industrial specification and evaluation, especially for drilling applications.

2. MATERIALS

Seven types of hard materials listed as BH1 – BH7, namely five types of engineering ceramics and two types of tungsten carbide (WC in abbreviation) varying cobalt contents were studied in this paper. In detail, BH1 is

[1] Corresponding author (J. Xu): tel. +49 5141-203-6669; fax: +49 5141-203-655; email: jing.xu@inteq.com

silicon carbide (SiC); BH2 is mainly silicon nitride (Si_3N_4) also containing aluminum oxide (Al_2O_3) and yttrium oxide (Y_2O_3) with the weight percentage of 90%Si_3N_4-6%Al_2O_3-4%Y_2O_3; BH3 is aluminum oxide (Al_2O_3); BH4 and BH5 are sintered YSZ-Al_2O_3 composite (YSZ is yttrium oxide stabilized zirconium oxide) with different Al_2O_3 percentage; BH6 is WC with 3%wt of cobalt (WC-3Co); BH7 is WC with 6%wt of cobalt (WC-6Co). Those materials are produced and supplied by different manufacturers.

Table 1 lists main physical, mechanical & thermal properties, and figure 1 shows the microstructure of the hard materials covered in this paper. It is found from table 1 that both strength, including flexural strength and compressive strength, and hardness of WC is superior over the other engineering ceramics. As is known, fracture toughness is the tradeoff for materials hardness. Therefore, WC does not behave the highest value of the fracture toughness, neither the hardness. However, the combined performance counting both toughness and hardness shows that WC behaves better than engineering ceramics. In detail, the hardness of BH6 is higher than BH2 & BH3 & BH4 & BH5, while it is 16% lower than BH1. Meantime, BH6 shows 170% higher fracture toughness value than BH1. Although BH4 has the highest toughness value (9.0 MPa·$m^{1/2}$), its hardness is very weak, i.e. 1200, comparing with BH6 with hardness 2020. Among two tungsten carbides with different cobalt content, increasing cobalt percentage causes a great drop of the hardness value while slightly improves the facture toughness.

Table 1 Physical Property, Mechanical Property, and Thermal Property of Engineering Ceramics and Tungsten Carbide

Properties	Units	BH1	BH2	BH3	BH4	BH5	BH6	BH7
Density	g/cm³	3.13	3.21	3.93	6.05	5.50	15.25	14.80
Open Porosity	%	3	0	0	0	0	0	0
Mean Grain Size	Micrometer	1-10	1-10	10	less than 1	less than 1	less than 1	less than 1
Flexural Strength	MPa (psi x 10^3)	550 (80)	760 (110)	350 (51)	900 (131)	1200 (174)	2300 (334)	3500 (508)
Elastic Modulus	GPa (psi x 10^6)	400 (58)	320 (46)	380 (55)	200 (29)	240 (35)	670 (97)	630 (91)
Compressive Strength	MPa (psi x 10^3)	3000 (435)	3000 (435)	3500 (508)	2500 (363)	2800 (406)	6700 (972)	7200 (1044)
Hardness	HV10	2350	1600	2000	1200	1500	2020	1820
Fracture Toughness	MPa $m^{1/2}$	3.0	8.0	4.0	9.0	8.0	8.1	8.2
Thermal Conductivity	W/m °K	125	30	30	2	5	95	90

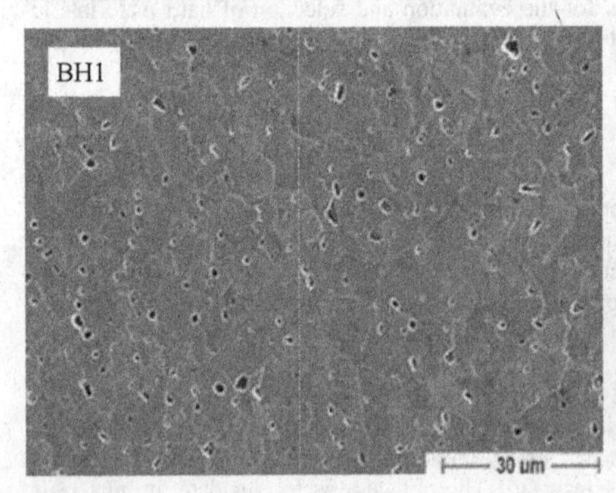

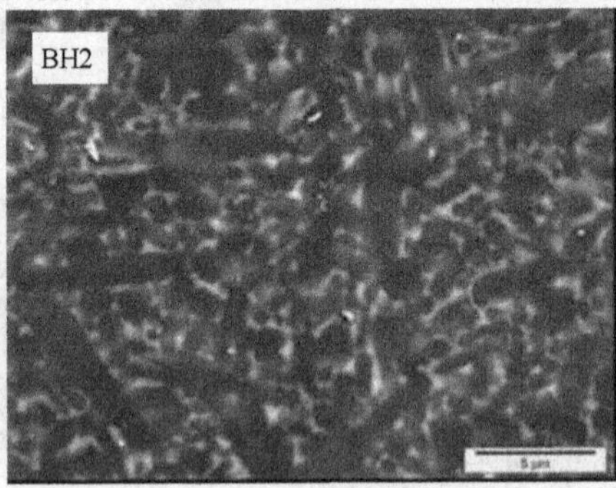

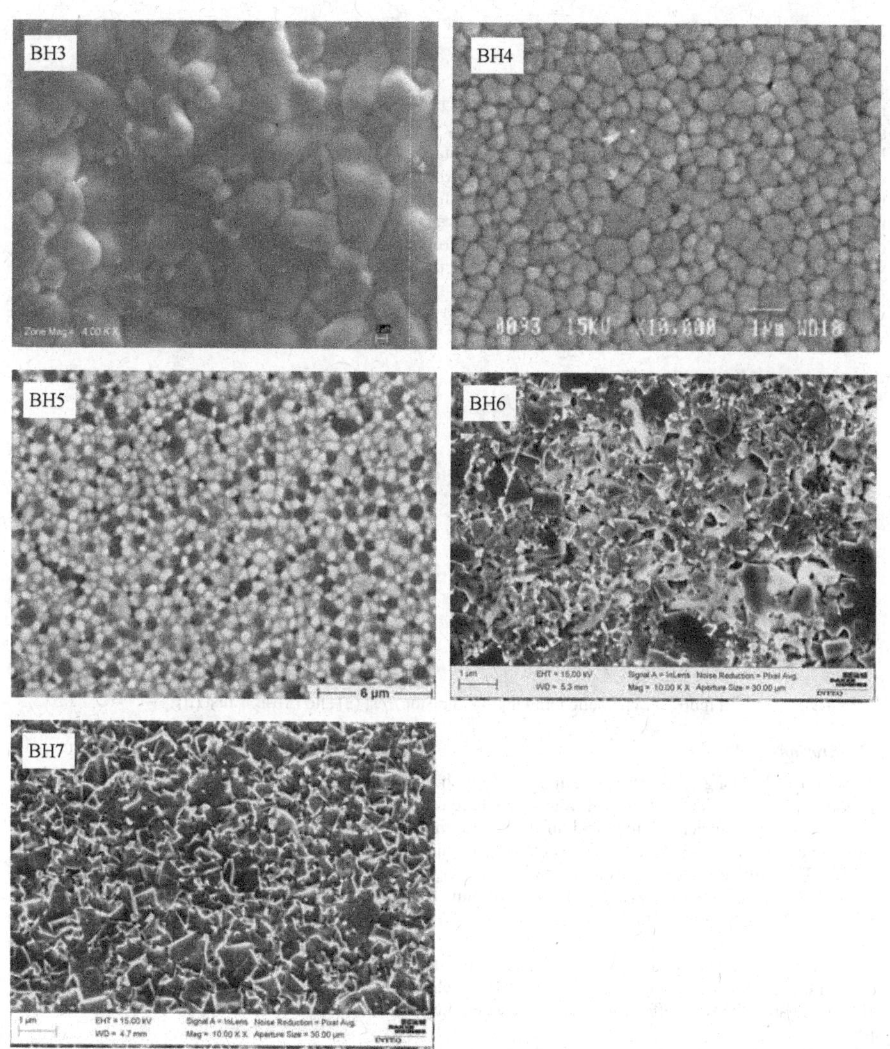

Figure 1 Microstructure of Hard Materials BH1 ~ BH7

3. EXPERIMENTAL METHODS

3.1 Abrasion

The specimens were 53.5 mm in length, 18.0 mm in width and 7.0 mm thick and were abraded at room temperature in an apparatus similar to ASTM G105-02. The schematic drawing of the abrasion testing set-up is shown in figure 2 (a). It consists basically of a vertically positioned steel wheel with an abrading Nitrile rubber layer, which rotates with its circumferential edge in contact with a horizontally mounted test sample on a lever arm. The test specimen is pressed against the rotating wheel by means of the lever arm while the grid abrades the test surface. The rotation of the wheel is such that stirring paddles on both sides agitate the abrasive slurry through which it passes to provide grit particles to be dragged across the contact face in the direction of wheel rotation. The abrasive mass flow is controlled by a feed of slurry mixed with alumina and water, and the velocity of abrasive movement through adjustment of the wheel rotational speed. The load was applied by adding weight onto the specimen holder directly above the area of interaction. All the materials were abraded with multi-edged alumina particles in the size range 500~700 micrometer with an applied load of 200 N and a wheel speed of 245 rpm or 2.28 m/s.

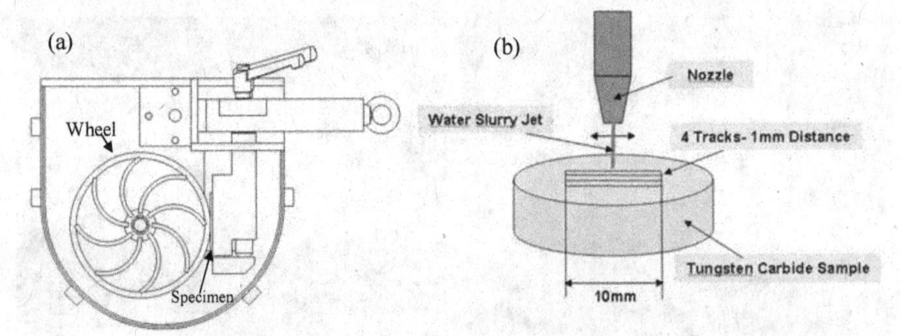

Figure 2 Schematic Drawing of Abrasion Test (a) and Erosion Test (b)

3.2 Erosion

The schematic drawing of the erosion testing set-up is shown in figure 2 (b). In the high speed particle/water jet erosion test, silica particles mixed with water were used to impact a target from a nozzle. The specimens used in this test are of diameter 25 mm and thickness 15 mm. The nozzle diameter is 0.3 mm. The angle of impingement was held constant at 45° and the erodent injection rate 6g/s. The blasting time is controlled as 600 seconds. The particle velocity was controlled by adjusting the water pressure to 3000 Bar. The focus diameter is 1.0 mm. The distance from nozzle to sample is 10 mm. The erodents, used for all erosion tests covered in this paper, are selected as silica with the average size of 280 micrometer.

Some erosion tests also used 75° impact angle or 60° to measure hard materials erosion performance. It is reported that the wear resistance values under 75° are close to that at lower impact angles such as 45°. [3, 4] In order to enhance the wear effect of silica particles eroding against the tested materials, 45° was chosen in this study.

4. ABRASION EROSION RESULTS

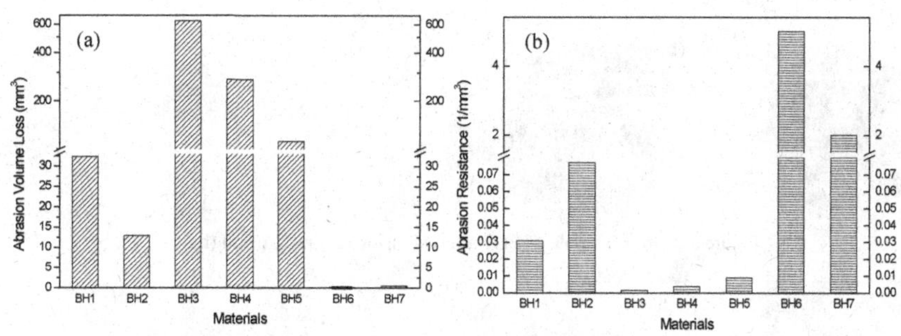

Figure 3 Abrasive Volume Loss (a) and Abrasive Wear Resistance (b) for the Tested Materials.

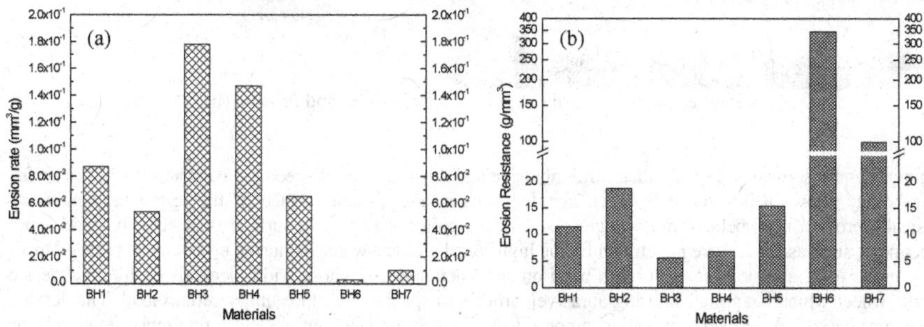

Figure 4 Erosive Rate (a) and Erosive Wear Resistance (b) for the Tested Materials.

The abrasive and erosive wear resistance results are summarized in the figures 3-4. In all abrasion tests and erosion tests, the ceramics showed higher wear rate than tungsten carbides, regardless of the various ceramics type and different cobalt content functioned as binder for tungsten carbide. BH3, namely aluminum oxide, showed in both abrasive wear tests and erosive wear tests the highest wear rates. In detail, BH6 was used as a reference to evaluate hard materials wear resistance, since BH6 held the lowest wear rate. BH6 showed twice the abrasion resistance than BH7, 65 times better than BH2, 160 times better than BH1, 560 times and 1360 times respectively than BH4 and BH5, and 3160 times than BH3. Meanwhile, to evaluate the erosion performance, BH6 showed three times the erosion resistance than BH7, 18 times better than BH2, 30 times better than BH1, 22 times and 50 times respectively than BH4 and BH5, and 61 times than BH3. The difference among materials was higher for the abrasion tests; a factor of three orders for abrasion and a factor of one order for erosion. The ranking among the tested 7 hard materials based on the abrasion resistance is nearly in accordance with that based on the erosion resistance, although abrasion and erosion are different in the manner for evaluating the wear resistance of hard materials. One exception of abrasion-erosion ranking result is that BH1 is superior to BH5 in abrasion resistance while BH1 is inferior to BH5 in erosion resistance. For TUNGSTEN CARBIDE CERMET materials of BH6 and BH7, abrasive wear and erosive wear depends on grain size, volume fraction of tungsten carbide particles in the composites.

5. MORPHOLOGY AND STRUCTURE EXAMINATION

5.1 Macroscopic Morphology

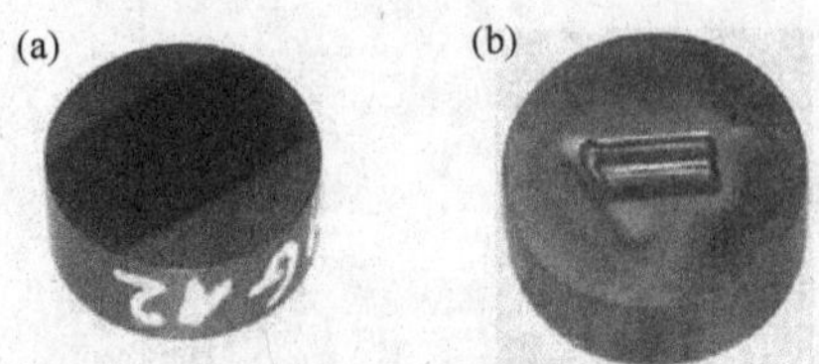

Figure 5 Hard Materials of BH7 after Abrasion (a) and Erosion (b)

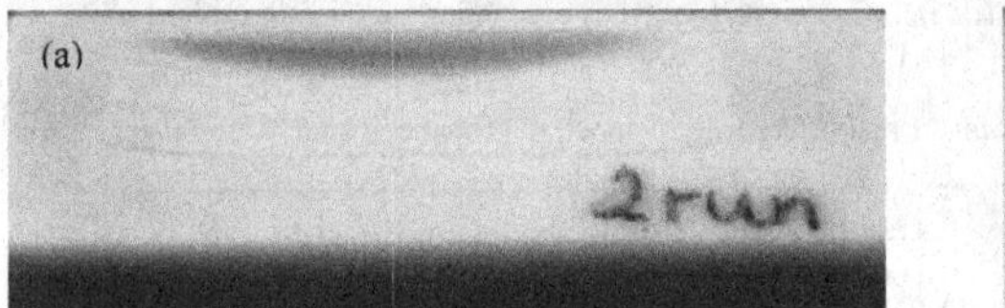

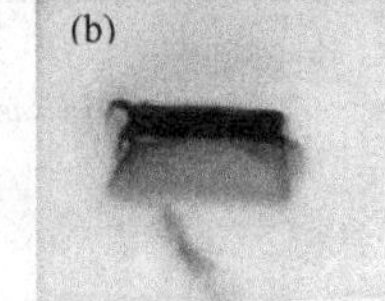

Figure 6 Hard Material BH3 after Abrasion (a) and Erosion (b)

Figures 5 and 6 show materials surfaces after abrasive and erosive wear of specimen BH7 and BH3. The surface morphology shows a difference between abraded surface and eroded surface of the tested hard materials. Basically erosion tests behave more aggressive in damaging the specimen surface than abrasion tests. Some specimens, such as BH3, were penetrated by the high speed erodent-water jet during erosion test, referred to fig. 6(b). In contrast, abrasion leaves a worn band on specimen surface, whose width depends on the thickness of rubber wheel which is applied to bring abrasive particles to specimen surface under certain load. The depth of the worn band left on surface varies among tested materials. Higher abrasion resistance materials are characterized by a shallower worn band. As discussed above, BH7 shows a much better wear resistance than BH3, which reveals the highest wear rate. Abrasion wear leaves a deep worn band in the middle of BH3 specimen surface while that of BH7 looks very shallow, referred to figs. 5(a) and 6(a). On the other hand, erosion wear surface shows four parallel grooves, if the specimen is not permeated by the particle-water jet. The local area surrounding eroded grooves is either partially deformed, see fig. 5(b) of material BH7, or partially chipped off, see fig. 6(b) of material BH3.

5.2 Microscopic Structure

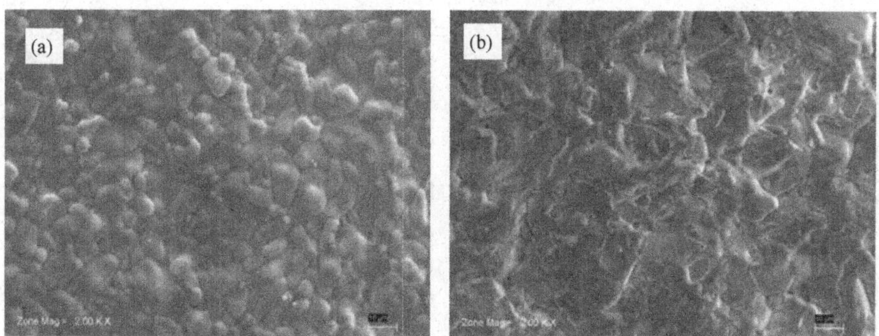

Figure 7 Microstructure of Hard Materials BH3 before (a) and after (b) Abrasion

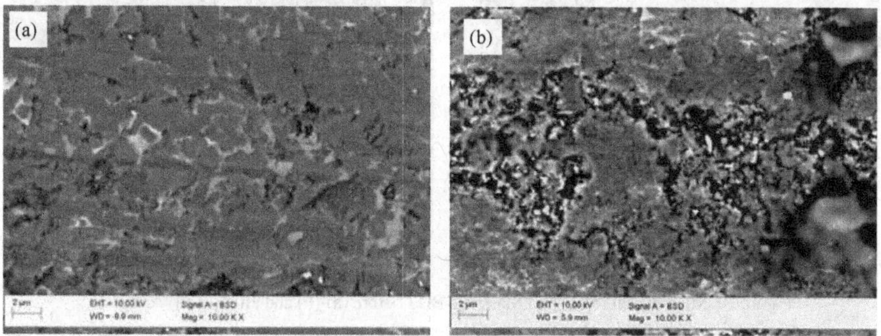

Figure 8 Microstructure of Hard Materials BH2 before (a) and after (b) Abrasion

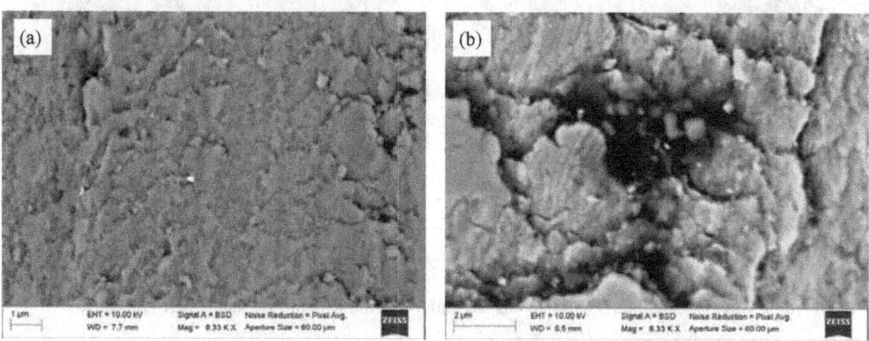

Figure 9 Microstructure of Hard Materials BH4 before (a) and after (b) Abrasion

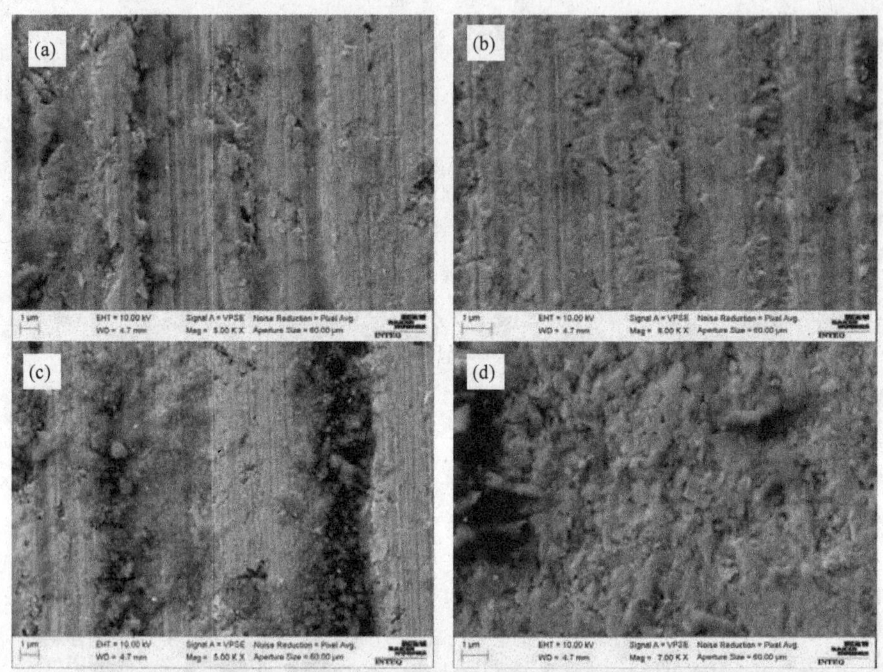

Figure 10 Microstructure of Hard Materials BH7 before (a)-(b) and after (c)-(d) Abrasion

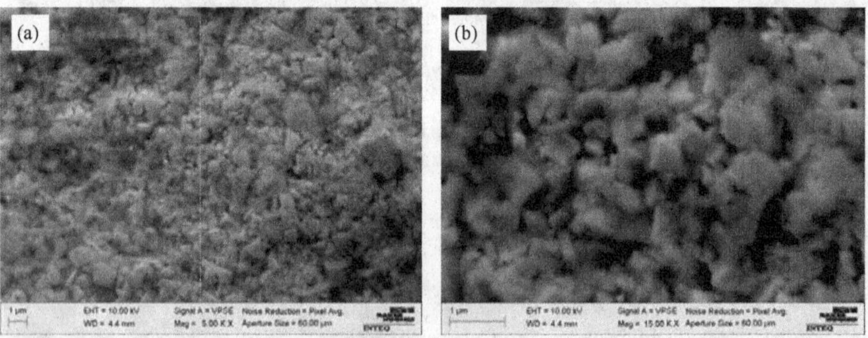

Figure 11 Microstructure of Hard Materials BH7 after Erosion with low (a) and high (b) magnification

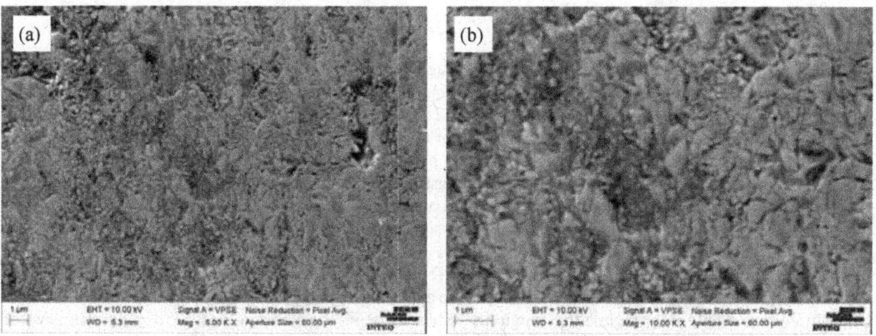

Figure 12 Microstructure of Hard Materials BH6 after Abrasion

The erosion of material BH6 and BH7 occurs through surface deformation of the bulk solid followed by fracture and the loss of small volumes of the tungsten carbide with binder. The passage of the erosive causes plastic deformation of the surface which results in the formation of grooves with materials piling up at the groove edges and at the head of the wear track, referred to fig. 5(b). SEM is applied to examine the local impact sites. Take material BH7 as example, the morphology after erosion is shown in fig. 5(b). After zoom in the impact sites, namely the deep grooves, it reveals that during erodent impact, material is deformed and the texture is not as dense as before. Some tungsten carbide particles are chipped off, due to the binder worn out. Removal of the binder phase occurs prior to any hard particle loss. The overall eroded surface shows ragged microstructure with submicron particles loosely piled, illustrated in fig. 11.

Conversely, the abraded surface of material BH6 and BH7 is relatively smooth and featureless, see figs. 10 and 12. Little penetration of the surface by alumina is observed. The cutting action of the abrasive leads to the loss of small volumes of materials. There is no evidence of the fragmentation of tungsten carbide grains due to the ultrafine grain size. Removal of the binder and ultrafine grains occurs partially. Some individual grains are torn away, but the amount of tungsten carbide particles lost during abrasion is much less than that lost during erosion, see figs 11(a) and 12(a). The abrasive surface, shown in fig 12(a), is smoother than the erosive surface, shown in fig 11(a).

During wet sand/rubber wheel abrasive tests, the coarse-grained materials, namely BH1, and BH3, the worn surfaces suggest that the grain pull-out phenomenon is the dominant mechanism determining the wear rate. Take material BH3 as an example, fig. 7 shows the comparison between the smooth surface before abrasion and rough surface after abrasion. The orderly sintered grain structure is smeared off and some aluminum oxide particles are pulled out, leaving sharp edges of surrounding grains remained, see fig. 7(b).

For tested materials holding low hardness, namely BH2, BH4, and BH5, the microstructure of abrasive surface indicates the wear mechanism in a predominantly brittle fashion in which intergranular fracture along the grain boundary between either same or different type of oxides and some transgranular fracture of the big size grains predominates. This leads to fragmentation and the loss of single grains together with spalling and the removal of large volumes of material and a high degree of surface damage, see figs. 8 and 9.

6. DISCUSSION

6.1 Wear Mechanism

The wear mechanism for cemented ceramics and tungsten carbide cermet is studied. For engineering ceramics, the surface shearing, caused by wear, displaces the structural alignment of cemented ceramic particles, resulting in some hard particles extruding towards the surface. The Following abrasion attacks locally on the extruded part, leading to either micro-fracturing or pulling-out of the hard particles. Another pattern is that big size hard particles are micro-fractured first, then those fractured particles being dragged out by partially losing bonding area. On the other hand, abrasion grooving attacks tungsten carbide cermet by damaging the binder phase first, which extrudes towards the surface during abrasive loading and shearing. The carbide grains thus lose their binder phase support and slip, fracture or fall out of the surface.

In abrasion, some hard materials show a tough wear regime with high wear while others show a mild wear regime with low wear, depending on materials type, the applied abrasion conditions. Some materials even show a transition from a mild wear regime to a tough one, when variating the abrasion load, the size and hardness of the abrading particles. [5] The transition can be coupled to a change in wear mechanism from ductile micro-cutting to brittle micro-fracturing. [6, 7]

Allen et al. [8] and Davidge et al. [9]suggested that at grain sizes larger than 2 micrometers, the dominant wear mechanism appears to be grain micro-fracture, which leads to fractured grain pull-outs in the end. Regarding the WC-Co cermet discussed in this paper with ultra-fine grain structure, namely sub-micrometer, a general observation about the abrasion behavior is that they undergo plastic deformation and micro-cutting to a much higher extent than ceramics.

Abrasion and erosion are different in the manner for evaluating the wear resistance of hard materials. In the case of abrasion, the abrasive particles are dragged over the surface of specimen at low velocity while in erosion the velocity of the erodent impacting is much greater, resulting in loading and strain rates which are significantly different in two testing forms. However, the results show a vast similarity in the way of hard materials responding to abrasion and erosion. Deformation, fragmentation and displacement of grains occurs during both abrasion and erosion which combined with the deformation, extrusion and fracture of the binder phase, is leading to the removal of the volume of the hard materials.

6.2 Size Effect of Grains and Particles

Materials removal in wear tests is a local resistance to deformation. Therefore, the size of the individual surface deformation as related to the dimensions of the materials microstructure is an important parameter for the prediction of wear resistance of multiphase materials such as cemented engineering ceramics and tungsten carbide cermet. Fine grains reduce the rate of wear, because of the delays in crack propagation at multiple boundary junctions for readjustment of the direction of crack propagation. The effect of particles may be more dominant than that of grain size because of their dual role of reducing grain size and modifying crack paths. It is believed that under most of conditions involving wear, ultrafine materials display a strong tendency to ductile behavior with all the attendant positive benefits of high wear resistance. However, the grain size is not the only factor determining the wear resistance. Actually, it is proved that wear resistance of the worst performing fine grained material is not substantially better than the wear resistance of the coarse grained materials. [8]

6.3 Comparison of Abrasion and Erosion

The hardness of abrasive particles, which applied for abrasion tests, affects the wear mechanism. [10] Alumina, which holds a high hardness value among the studied hard materials, was applied as abrasive particles. This selection of the hard particles maximizes the deformation and fracture of the hard material grains and suppress and ductile deformation response. However, it is not practical to use alumina particles as erodents, since the high hardness of erodents and high jet rates applied during erosion causes severe fracture damage to each studied materials type, which makes it difficult to measure the erosive wear as well as to rank wear resistance among those hard materials. Feng and Sapate studied the effect of erodent hardness on the wear performance of WC-Co and found that tungsten carbide grains eroded with high hardness particles showed a significant degree of fracture while no facture was found in the tungsten carbide eroded with low hardness silica particles which lead to higher wear rates. [11, 12]

Erosion response depends on the scale of the impact zone relative to the microstructure. [13, 14] It is reported that when the impact crater encompasses less than 10WC grains material removal occurs mainly through the brittle response of the individual WC grains while if the impact zone encompasses more than 10 grains the material response consists mainly of bulk ductile deformation. [8] Since the erosion tests in this paper applied the impact zone 1 mm, which covers far more than 10 grains regardless of the studied hard materials variety, the damage mainly comes from the bulk ductile deformation.

6.4 Comparison of Engineering Ceramics and Tungsten Carbide Cermet

Tungsten carbide cermet show more restraint to wear due to the effect of particles. They dissipate the energy of fracture through various phenomena such as deflection of the cracks into the grain. It was reported under TEM observation that this phenomena occur even inside grains as small as about 1 micrometer. Bridging and weak toughening were found with very short distances of the order about 120 nm. For a given area there are more bridging sites in the composites than that in bulk ceramics, which contributes to performance of better wear resistance. [15]

For tungsten carbide cermet, the binder type, amount, and the alignment of the particles with binder also play an important role for the performance in wear resistance. For example, for about the same grain size, WC-3Co behaves higher abrasive-erosive resistance than WC-6Co.

7. CONCLUSIONS

The wear behaviors of seven types of hard materials used in oil and gas drilling application have been systematically investigated at two different wear conditions, namely the wet sand/rubber wheel abrasive wear and the high speed particle/water jet erosive wear.

Under the tested circumstances tungsten carbide cermet exhibits a higher wear resistance than engineering ceramics such as alumina, zirconia, silicon carbide, and silicon nitride. Abrasive wear testing results are in general in accordance with erosive wear testing results, regarding to the wear resistance ranking.

Deformation, fragmentation and displacement of grains occur during both abrasion and erosion. Depending on the wearing condition and hard materials structure and composition, the wear mechanism could be dominant by either brittle grain micro-fracturing or ductile grain micro-cutting or by a combination of the above two.

Size of grains and particles plays an important role to the wear behavior of the hard materials. Fine grains reduce the rate of wear. Under most of the conditions involving wear, ultrafine materials display a strong tendency to ductile behavior with all the attendant positive benefits of high wear resistance.

For tungsten carbide cermet, the binder type, amount, and the alignment of the particles with binder also play important role for the performance of wear resistance. Results show that the higher the binder content was, the lower was the wear resistance.

ACKNOWLEDGEMENTS

The authors acknowledge the contributions of Dominik Steinhoff at University of Applied Sciences, Hannover, in support of the abrasion tester described in this paper. The efforts of Joachim Oppelt from Baker Hughes INTEQ in hard materials application in oil drilling field are also appreciated. We also thank Volker Krueger from Baker Hughes INTEQ, and H.-T. Lin from Oak Ridge National Laboratory in reviewing the manuscript before submission.

REFERENCES:

1 Shan-Ping Lu, Oh-Yang Kwon, Yi Guo, Wear behavior of brazed WC/NiCrBSi(Co) composite coatings, Wear 254(2003), pp. 421-428
2 U. Beste, L. Hammerström, H. Engqvist, S. Rimlinger, S. Jacobson, Particle erosion of cemented carbides with low Co content, Wear 250(2001), pp. 809-817
3 K. Jia, T.E. Fischer, Wear 200(1996), pp. 206-214
4 Jong Jip Kim, Seong Khil Park, Solid-Particle erosion of hot-pressed silicon carbide and SiC-TiB2 composite, J. Materials Science Letters, 16(1997), pp. 821-823
5 K. Zum-Gahr, Microstructure and Wear of Materials (Elsevier, Amsterdam, 987), pp. 304-313
6 H. Engqvist, N. Axén, S. Hogmark, Resistance of a binderless cemented carbide to abrasion and particle erosion, Tribology Letters 4(1998), pp. 251-258
7 M.G. Gee, A. Gant, B. Roebuck, Wear mechanisms in abrasion and erosion of WC/Co and related harmetals, Wear 263(2007), pp. 137-148
8 C. Allen, M. Sheen, J. Williams; V.A. Pugsley, The wear of ultrafine WC-Co hard metals, Wear 250(2001), pp. 604-610
9 R.W. Davidge, P.C. Twigg, F.L. Riley, Journal of European Ceramics Society, 16(7)(1996), pp. 799-801
10 C.N. Machio, G. Akdogan, M.J. Witcomb, S.Luyckx, Performance of WC-VC-Co thermal spray coatings in abrasion and slurry erosion tests, Wear 258(2005), pp. 434-442
11 Z. Feng, A. Ball, Nordtrib 98, Proceedings of the 8th International Conference on Tribology, Vol. 1, DTI Tribology Centre, Aarhus, Denmark, 1998, pp. 15-26
12 S.G. Sapate, A.V. RamaRao, Erosive wear behavior or weld hardfacing high chromium cast irons effect of erodent particles, Tribology International 39(2006), pp206-212
13 R.A. Saravanan, M.K. Surappa, B.N. Pramila Bai, Erosion of A356 Al-SiCp composites due to multiple particle impact, Wear 202(1997), pp. 154-164
14 V.A. Pugsley, C. Allen, Microstructure/property relationships in the slurry erosion of tungsten carbide-cobalt, Wear 225-229(1999), pp. 1017-1024
15 C. C. Anya, Wet erosive wear of alumina and ist composites with SiC Nano-particles, Ceramics International 24(1998), pp. 533-542

THERMAL BARRIER COATINGS DEPOSITED BY THE FARADAYIC EPD PROCESS

Joseph Kell, Heather McCrabb
Faraday Technology, Inc.
Clayton, Ohio, USA

Binod Kumar
University of Dayton Research Institute
Dayton, Ohio, USA

ABSTRACT
Faraday Technology is developing a non-line of sight, electrically mediated electrophoretic deposition process (Faradayic EPD) for thermal barrier coatings for use in gas turbines. This process has been shown to produce uniform coatings through the application of pulse and pulse-reverse electric fields that, when compared to traditional EPD, have an increased coating uniformity and decreased surface roughness due to a decrease in hydrolysis and edge effects brought about by better control of the electric field during deposition. A yttria-stabilized zirconia (8% YSZ) coating was deposited from a suspension onto Ni-based superalloy substrates that had an MCrAlY bond coat. The samples were then subjected to a binder burnout and sintering process and subjected to a battery of tests, including surface roughness measurements, microstructural examination, thermal conductivity measurements, thermal cycling tests, and other appropriate tests of merit, to determine the feasibility for use in gas turbine engines. The primary objective of this research is to develop the EPD process to produce TBCs with properties consistent with similar deposition methods.

INTRODUCTION
Gas turbine engines present a significant challenge to designers and manufacturers due to the severe conditions present in their operating environments. The hot section of a turbine engine is especially hostile to materials that make up the turbine blades and vanes. The environment of the hot section consists of hot, corrosive gasses, usually contaminated with chlorides and sulfates, an enriched oxygen atmosphere, severe mechanical wear and erosion[1], and harsh thermo-mechanical conditions often including thermal cycling and thermal shock. Thermal barrier systems were developed as a cost effective way of protecting the components and materials within gas turbine engines from damage due to the severe environment. In typical thermal barrier systems, a bond coat (BC) is deposited onto the substrate material to perform two functions; it provides enhanced oxidation resistance and helps alleviate the stress that develops due to thermal expansion mismatch between the substrate (usually a Ni-based superalloy) and the TBC (often yttria-stabilized zirconia)[2,3]. The TBC is then placed on top of the bond-coated substrate. The TBC serves to protect the system from the mechanical, oxidative, and thermal environments by inducing a steep thermal gradient through the coating system, resulting in the substrate experiencing lower temperatures and therefore having improved properties. For the TBC to provide this protection, certain material requirements are needed, specifically chemical inertness, a high melting point, low thermal conductivity, a low sintering rate, and a thermal expansion coefficient similar to the base substrate[4].

Yttria-stabilized zirconia (YSZ) is the most commonly used TBC material in today's gas turbine engines. The material is largely chemically inert, will not oxidize further and when applied onto the bond coat, provides a degree of corrosion protection. In addition, YSZ performs well when subjected to thermal cycling and thermal shock, largely due to its relatively high thermal expansion coefficient and melting point. It has a low thermal conductivity, which further helps protect the bond coat and substrate against the high temperatures encountered within gas turbine engines[5]. In industry, YSZ TBCs are typically applied by either electron beam physical vapor deposition (EB-PVD)[6,7] or air plasma spray (APS)[8,9]. Both of these deposition methods have limitations, which are often contradictory. APS coatings have a low thermal conductivity and can be produced quickly and economically, however, they are noted for having thickness variations and spallation problems[10,11]. EB-PVD produces films that exhibit good wear and spallation resistance but are generally slower and more expensive to produce and have a higher as-sprayed thermal conductivity than their APS counterparts[7,8,11]. However, it may be possible to address these problems through the electrophoretic deposition (EPD) of TBC materials.

Conventional EPD, or direct current (DC) EPD, consists of immersing 2 electrodes, one of which is the material to be coated, into a stable suspension of charged deposition particles and applying a constant voltage across them, creating an electric field that then migrates the particles to the oppositely charged electrode (substrate)[12]. The field is capable of readily wrapping itself around the electrodes, resulting in a largely non-line[13] of sight process. In addition, DC EPD has other distinct advantages, including fast deposition rates (> 5 mm/min)[14], the ability to coat complex shapes uniformly, low levels of coating contamination, a reduction of material waste encountered in spraying or non-directional coating methods, and relatively simple deposition equipment. Through the application of electrically mediated waveforms, a process known as the *Faradayic* EPD process, many of the difficulties of DC EPD, such as the thickness non-uniformities, hydrolysis, and other electrochemically related difficulties can be overcome[15]. Previous work at Faraday has demonstrated that electrically mediated waveforms (including pulse, pulse-reverse, and more complex waveforms) can substantially modify the deposition related characteristics of electrodeposited coatings[16,17,18,19] and anodically finished surfaces[20] through enhanced control of the electric field. These same principals are applied to the EPD process in an attempt to improve the coating technique.

EXPERIMENTAL

Suspensions consisting of YSZ particles (8 wt% Y_2O_3-stabilized ZrO_2, submicron powder, Sigma Aldrich, 30-60% wt% of suspension), polyvinyl alcohol (86-89% hydrolyzed, dissolved in water, <5% v/v of suspension) (PVA), and poly(diallyldimethylammonium chloride, <5% v/v of suspension) (PDDA) in ethanol (balance) (EtOH) were made for EPD. PVA was used as a binding agent and PDDA as the cationic dispersant. YSZ, EtOH, and PDDA were mixed and ultrasonicated together to create a stable suspension for coating. The PVA was then added last to prevent competitive adsorption between the binding and charging agents and the suspension was again mixed, ultrasonicated and aged overnight. For depositions, the pH varied from 4 to 11.

EPD experiments were performed in a glass electrophoresis cell. The iridium oxide coated titanium anode and bond coated nickel-based superalloy cathode were fixed one cm apart. The substrate holders, which were manufactured in-house, were made from glass reinforced PTFE to avoid contamination issues and to prevent warping of the holder during the drying step. The cathode and anode were kept in the vertical plane during deposition and the particles were deposited onto the face of the cathode. The suspensions were mechanically agitated immediately before each EPD experiment to ensure uniform particle distribution for the deposition. The cathodic substrates consisted of NiCoCrAlY bond coated IN939 buttons with a nominal diameter of 2.54 cm. Basic EPD conditions, bath properties, and substrate properties are outlined in Table I.

For this work, forward pulsed waveforms of varying duty cycles and frequencies were employed with an electric field of 50 V/cm. The duty cycle (γ) is defined as the percentage of time the electric field is turned on and the frequency (f) is the inverse of the sum of the electric field on-time and off-time for a single pulse. The basic waveform for the pulsed EPD depositions is shown in Figure 1. The frequency and duty cycle were chosen based off of previous optimization work carried out at 50 V[17,18]. To allow for comparison with prior EPD experiments, the total deposition time for the *Faradayic* EPD samples was calculated by using the following equation:

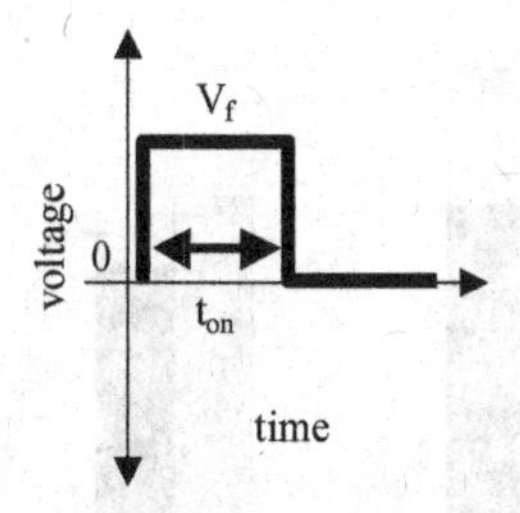

Figure 1: Basic waveform for the *Faradayic* pulsed EPD experiments. V_f is the peak voltage of the forward pulse and t_{on} is the time of a single pulse. t_o=Ót_{on}

$$E_a*t_d*d_c = E_a*t_o = 900 \text{ Volts*sec.*cm}^{-1} \quad (1)$$

where E_a is applied electric field in V/cm, t_d is the total deposition time in seconds, d_c is the duty cycle (unitless) and t_o is the sum of the total time that the electric field is on during the experiment in seconds (t_o = 180 s was used in previous work). The depositions were performed using voltage control. Typical deposition and substrate properties are shown in Table I. After the deposition, the samples were dried for a few hours at ~95°C to drive off the solvent (i.e. EtOH) and immediately cooled in a desiccator.

Table I: Typical Substrate, Bath, and EPD conditions for the *Faradayic* EPD of TBCs

Substrate Properties		Bath and Deposition Properties	
Base Material	IN939	Electrode Distance	1 cm
Bond Coat	NiCoCrAlY	Binder	PVA
Diameter	~2.54 cm	Charging Agent	PDDA
Thickness	0.35 cm	Electric Field	50 V cm^{-1}
Bond Coat R_a	10.84 ± 0.56 µm	Current Density	2.7 mA cm^{-2}
		Deposition Rate (50 V)	29.5 ± 5.7 mg cm^{-2} min^{-1}

Microhardness, thermal cycling, energy dispersive spectroscopy studies done on a scanning electron microscope (SEM/EDS), and a comparison of the cross-sectional microstructure of various samples were performed. Thermal cycling was done on a samples of low γ and high f sintered at 1092 °C with cycles consisting of 1 hour heating followed by air cooling to room temperature at 1000 °C and 1092 °C. These temperatures were chosen to give good comparison to literature and to allow for initial, baseline behavior for the development of the process. Samples were cycled for ~100 hours to allow comparisons between them. Photographs of the sample were taken every few cycles to determine the degree of degradation of the sample.

Samples for cross sectional microstructural examination were first encased in epoxy using Buehler EpoHeat™ epoxy. The samples were then cross-sectioned using a general-purpose diamond blade and remounted. The remounted samples were ground with 320-grit SiC paper and polished with 9 µm, 3 µm, and 1 µm diamond paste and a mixture of Buehler MASTERMET and MASTERPREP polishing suspensions. The samples were then imaged with both an optical microscope and SEM, and tested for microhardness. Microhardness was performed on three samples; a DC EPD sample sintered at 1000 °C for 1 hour, a *Faradayic* EPD sample processed with a medium γ and high f sintered at 1092 °C for 2 hours, and a *Faradayic* EPD sample processed with a high γ and high f sintered at 1092 °C for

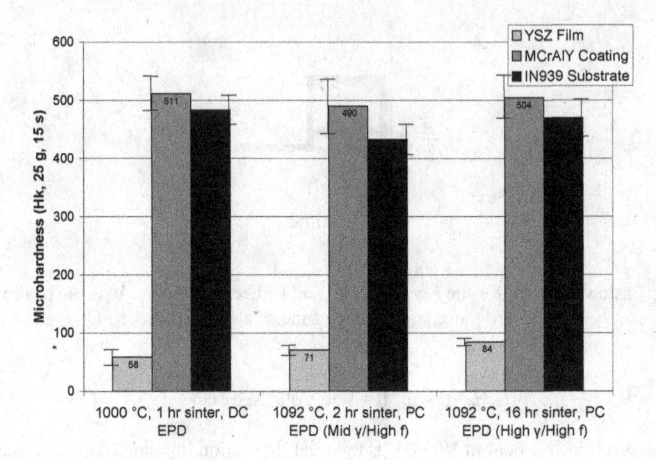

Figure 2: Microhardness of 3 samples processed by EPD. Error bars represent one standard deviation.

16 hours. Samples were measured using a Knoop indenter with a 25g load and a test time of 15s. Indentation width was then converted to hardness. Five tests on each section of each specimen were performed and averaged together. SEM/EDS was performed on a LEI SEM with an Ametek EDAX detector. The sample was coated with carbon prior to imaging to minimize charging of the ceramic TBC.

RESULTS AND DISCUSSION

Microhardness

Figure 1 shows the results of the microhardness tests of the IN939 substrate, the MCrAlY bond coat, and the YSZ TBC film. Overall, the hardness of the BC and substrate were similar to accepted values while the YSZ had a very low hardness value due to the high levels of porosity. With respect to sintering temperature, there does not appear to be a significant change in the hardness of the bond coat and underlying substrate, suggesting that the sintering is not negatively impacting the materials underneath the TBC. In addition, the hardness of the TBC film appears to be increasing with increasing temperature and with increasing time. This is important as the optimal sintering temperature is not yet finalized and further research will look at continuing optimization of the sintering conditions. In addition, although it is believed that changing the EPD parameters will not significantly affect the hardness of the YSZ film, further tests are expected to be performed to confirm this.

EDS

EDS was performed on a sample with a medium γ and low f to analyze the BC-TBC interface region. Figure 3 shows the region of the sample that was investigated. It appears from both the picture and the trace that the thermally grown oxide (TGO) that forms between the BC and TBC consists of two distinct phases, an Al-Zr-rich oxide phase and an Al-rich oxide phase. This may be due to the TGO growing into the TBC due to the high porosity of the TBC and may be potentially advantageous from an adhesion point of view, although further tests will need to be done to determine if this is actually the case. The TGO is approximately 2 μm thick, with each layer being approximately 1 μm wide. As expected, Ni does not have a significant presence within the TGO and the oxide content

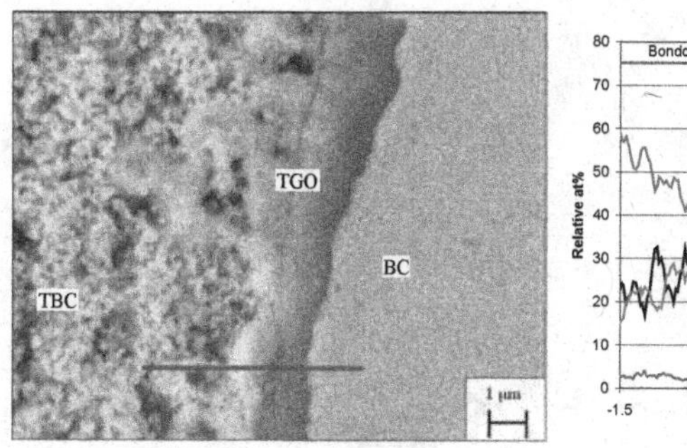

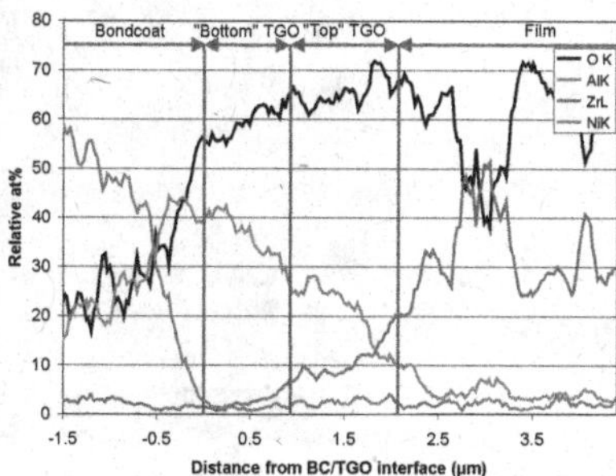

Figure 3: (left) Typical region investigated in the SEM/EDS investigation. Horizontal line shows region scanned to determine compositions of layers and the diffusion of species throughout the TBC, TGO, and BC. (right) Linear scan of the horizontal line in the picture on the left showing major constituents of the various layers in at%. The TGO that is present after sintering is shown as two separate layers (which is also evident in the picture as well), an Al-Zr-O phase near the TBC and an Al-O phase near the BC. The BC is negative on the scale and the TGO/TBC is positive.

drops off rapidly within the BC. The varying amounts of Zr and O content in the TBC film are believed to be coming from a combination of porosity and microstructure effects as well as error related to the low counts used in the EDS scan. As the Y content was very low and would likely be observed in all phases, it was left out in the interest of high accuracy in detecting the other elements.

Thermal Cycling

Two samples were submitted for thermal cycling, both of which were low γ/high *f* samples sintered at 1092 °C for 4 hours. The thermal cycling tests were identical except one sample was subjected to 1000 °C max temperature and the second was subjected to 1092 °C max temperature. Both samples were subjected to their respective temperatures for 1 hour and allowed to cool to RT in air. This was done as this test more closely approximates the end use conditions of the material system. Figure 4 shows before and after images of the sample during the thermal cycling test. Overall, the

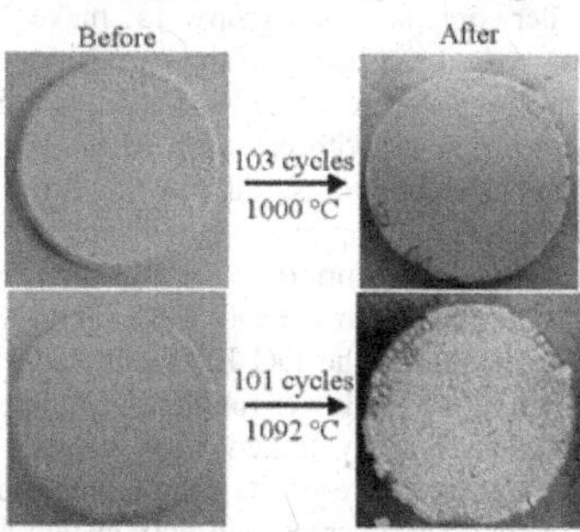

Figure 4: Photographs of samples before and after undergoing thermal cycling. Several black flakes are on 1092 °C sample and are believed to be oxide from the exposed metal of the sample. Spallation always started from the edges and proceeded inwards.

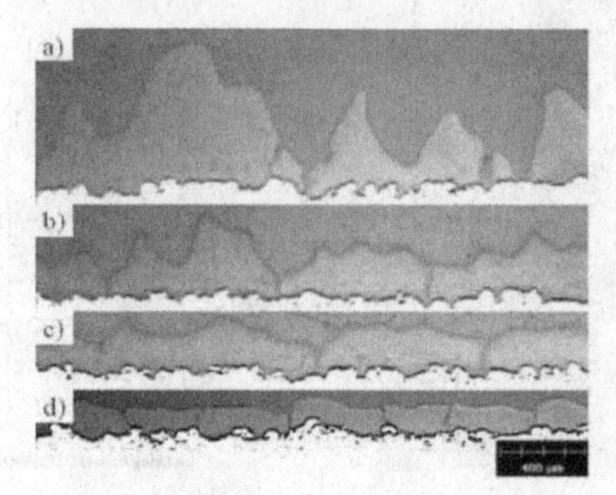

Figure 5: Cross-section micrographs of (a) DC-EPD sample (295 μm) compared with Faradayic EPD samples with (notation as γ/f) (b) high/high (185 μm), (c) medium/high (160 μm), (d) low/high (112 μm).

samples survived fairly well, showing only 2.9% spallation of the TBC film after 103 cycles at 1000 °C and 7.4% spallation after 101 thermal cycles at 1092 °C. However, the spallation that was observed occurred on the edges of the sample and did not occur on the interior of the sample. As the samples do not have a bond coat that wraps around the edges, it is expected that the edges would spall preferentially to the bulk and therefore this test may exaggerate the spallation that would have occurred if the samples were uniformly bond coated. Further thermal cycling work, thermal cycling of commercially prepared samples for comparison, and cross sections of the thermally cycled samples are ongoing.

Microstructure

In previous work[18], *Faradayic* EPD samples were sintered to determine if the use of a pulsed electric field in the EPD process would have positive effects on the properties of the YSZ coating. Prior work examined differences between samples processed with medium γ and high γ samples and varying f. In this work, low γ samples with high f are compared to medium and high γ samples processed at high f and to a conventional DC EPD sample. The samples were sintered prior to cross sectioning and were observed under optical microscopy to make comparisons between the microstructure of the *Faradayic* EPD coated samples and the conventional DC EPD coated samples (Figure 5). The samples used for these tests were coated using higher frequencies as this tends to produce films that appear to have less problems with cracking when compared to samples that were processed with low frequencies. It is not entirely understood why this is the case, however one possible explanation is the release of residual stresses built up in the film during deposition, a common problem seen in electroplating. All samples, with the exception of the low γ sample, showed the hydrolysis related parabolic groove defects that were noted in the DC EPD samples, which result in relatively high surface roughness values. However, unlike the DC EPD samples, these grooves almost never reached the bottom of the substrate, explaining the lower roughness values measured. The depth of the parabolic hydrolysis defects are therefore, heavily linked with γ in the microstructures. Higher γ approximate the DC conditions more closely and result in more extreme defects. However, thickness and deposition mass also scales with γ, and any decreases in duty cycle will lead to longer deposition times. It is also worth noting that the edge effects that are commonly a problem in electrochemical processes are also reduced for the *Faradayic* EPD samples. This results in an improvement to the thickness homogeneity at the edges of the samples when compared with the DC EPD samples and with samples of higher duty cycle. It is becoming more evident, with the inclusion of the relatively smooth low γ samples, that the through cracks appear to be located near local maxima in the bond coat roughness and are possibly shrinkage defects due to the rapid air-drying cycle. As such, it may be worthwhile to look at reducing the surface roughness of the BC layer and also in improving the drying cycle, especially if the through cracks are found to be a problem.

CONCLUSIONS

Several important conclusions were found in this work. First, the microhardness of the samples was taken and it was shown that the BC and IN939 substrate did not appear to be negatively effected from the sintering step. Also, the microhardness tends to support the fact that the film is very porous, almost having the consistency of foam. Second, EDS scans showed that the films were indeed of the desired composition and that the TBC-BC region is complex in the EPD samples, with possible ingrowth of the TGO layer into the TBC. Third, thermal cycling showed that the EPD films are capable of surviving 100+ cycles when subjected to 1092 °C, 1-hour cycles. Finally, the low γ samples were cross-sectioned and compared to other samples processed by DC EPD and *Faradayic* EPD and showed that they are indeed more uniform. Also, the low γ samples lend more evidence that mud cracking, that is evident in all of the bond coated EPD samples, might be due to the high roughness of the substrate. This mud cracking is a common trait of all wet, thick film processes and is evident immediately after the drying step of the process. Future work will focus on continuing to optimize the sintering process of the films, including higher temperature and alternative atmosphere sintering and other ways to preferentially heat the film, as opposed to the underlying material. Further work will also include exploration of the thermal cycling limits of the films to determine at what level the films can perform when compared to EB-PVD and APS processed films.

ACKNOWLEDGEMENTS

This material is based upon work supported by the Department of Energy under Grant No DE-FG02-05ER84202. Any opinions, findings, and conclusions or recommendations expressed in this material are those of the authors and do not necessarily reflect the views of the Department of Energy.

Faraday gratefully acknowledges Siemens Power Generation for supplying substrates for deposition experiments. In addition, Faraday would like to acknowledge Jitendra Kumar of UDRI for his assistance with the thermal cycling part of this work. Faraday also acknowledges Greg Wilt and Dr. Allen Jackson of Wright State University for their help with the collection of the microhardness data and SEM/EDS data, respectively.

[1]D. Wolfe, J. Singh, Functionally gradient ceramic/metallic coatings for gas turbine components by high-energy beams for high-temperature applications, *J. Mater. Sci.*, **33** (14), 3677-3692 (1998).

[2]D.R. Mumm, A.G. Evans, Failure of a Thermal Barrier System Due to a Cyclic Displacement Instability in the Thermally Grown Oxide, *Mat. Res. Soc. Symp. Proc.*, **645E**, M2.6.1-M2.6.6 (2001).

[3]C. Leyens, U. Schultz, M. Bartsch, M. Peters, R&D Status and Needs for Improved EB-PVD Thermal Barrier Coating Performance, *Mat. Res. Soc. Symp. Proc.*, **645E**, M10.1.1-M10.1.12 (2001).

[4]X.Q. Cao, R. Vassen, and D. Stoever, Ceramic materials for thermal barrier coatings, *J. Eur. Ceram. Soc.*, **24** (1), 1-10 (2004).

[5]T. Narita, S. Hayashi, L. Fengqun, K.Z. Thosin, The Role of Bond Coat in Advanced Thermal Barrier Coating, *Mat. Sc. Forum*, **502**, 99-104 (2005).

[6]D.D. Hass, P.A. Parrish, H.N.G. Wadley, Electron Beam Directed Vapor Deposition of Thermal Barrier Coatings, *J. Vac. Sci. Technol.*, **16** (6), 3396-3401 (1998).

[7]P. Hancock, M. Malik, *Materials for Advanced Power Engineer, Part I*, edited by D. Coutsouradis *et al.*, Kluwer Academic, Dordrecht, The Netherlands, 685-704 (1994).
[8]J.D. Vyas, K.L. Choy, Structural characterisation of thermal barrier coatings deposited using electrostatic spray assisted vapour deposition method, *Mat. Sc. Eng.*, **A277** (1-2), 206-212 (2000).
[9]H. Wang, R.B. Dinwiddie, Characterization of Thermal Barrier Coatings Using Thermal Methods, *Advanced Engineering Materials*, **3** (7), 465-468 (2001).
[10]W. Beele, G. Marijnissen, A. van Lieshout, The Evolution of Thermal Barrier Coatings-Status and Upcoming Solutions for Today's Key Issues, *Surf. & Coatings Tech.*, **120/121**, 61-67 (1999).
[11]E. Tzimas, H. Mullejans, S.D. Peteves, J. Bressers, W. Stamm, Failure of Thermal Barrier Coating Systems Under Cyclic Thermomechanical Loading, *Acta Materialia*, **48** (18-19), 4699-4707 (2000).
[12]T. Ishihara, K. Sato, Y. Takita, Electrophoretic Deposition of Y_2O_3-Stabilized ZrO_2 Electrolyte Films in Solid Oxide Fuel Cells, *J. Am. Ceram. Soc.*, **79** (4), 913-919 (1996).
REFERENCES
[13] T. Rosjford, D. Haught, D. Geiling, Cooperative Research and Development for Advanced Microturbine System, presented at the U.S. Department of Energy Distributed Energy Peer Review, Arlington, Va, December 13-15, 2005 (2005). Found at: http://sites.energetics.com/depeerreview05/pdfs/presentations/turbines/tu_b2-2.pdf
[14]O. Van der Biest, S. Put, G Anné, J. Vleugels, Electrophoretic Shaping of Free Standing Objects, in *Electrophoretic Deposition, Fundamentals and Applications*, ed. A.R. Boccaccini, O. Van der Biest, J.B. Talbot, ISBN 9781566773454, 60-69 (2002).
[15]H. McCrabb, M. Inman, E. Taylor, Faradayic Electrophoretic Deposition of Thermal Barrier Coatings, Poster Presented at *31st International Conference & Exposition on Advanced Ceramics and Composites*, Daytona Beach, Fl, USA, The American Ceramics Society, ICACC-S2-054-2007 (2007).
[16]E.J. Taylor, *et al.*, Electrically Mediated Plating of Semiconductor Substrates, Chip Scale Packages & High-density Interconnect PWBs, *Plating and Surf. Fin.*, **89**, 88, (May 2002).
[17]H.A. McCrabb, J.W. Kell, B. Kumar, Faradayic Process For Electrophoretic Deposition Of Thermal Barrier Coatings, presented at *32nd International Conference & Exposition on Advanced Ceramics and Composites (ICACC 2008)*, Dayton Beach, Fl, USA, The American Ceramics Society, ICACC-S8-013-2008 (2008).
[18]J.W. Kell, H.A. McCrabb, B. Kumar, Electrophoretic Deposition of Thermal Barrier Coatings by the Faradayic Process, *Proceedings of the Materials Science and Technology 2008 Conference: Enabling Surface Coating Systems: Science and Technology*, October 5-9, 2008, Pittsburgh, Pennsylvania, (Oct 2008).
[19]J.W. Kell, H.A. McCrabb, Faradayic Process for Electrophoretic Deposition of Thermal Barrier Coatings for use in Gas Turbine Engines, *Innovative Processing and Synthesis of Ceramics, Glasses, and Composites X: Ceramic Transactions, Volume 207*, ed. N.P. Bansal, ISBN:978-0-470-40845-2, To be published May 2009.
[20]J. J. Sun, *et al.*, Electrically Mediated Edge & Surface Finishing for Automotive, Aerospace & Medical Applications, *Plating and Surf. Fin.*, **89**, 94, (May 2002).

THE INFLUENCE OF THICKNESS ON THE PROPERTIES OF AIR PLASMA SPRAYED CERAMIC BLEND AT ROOM TEMPERATURE

Jason E. Hansel
Universal Technology Corporation
Dayton, Ohio

ABSTRACT

Previous research suggests that damping of metallic beams with hard ceramic coatings is proportional to the coating thickness. This indicates that the damping is a volume dependent material property. Since thickness variations are likely in an actual application, it is essential to understand the role that coating thickness plays in damping effectiveness. For this research, a series of tests were conducted using twelve substrate beams of 90 mil Ti-6Al-4V, coated with 3 mils of an air plasma sprayed NiCrAlY bond coat followed by a Titania-Alumina ceramic blend coating applied via air plasma spray in one of three thicknesses (5 mil, 10 mil and 15 mil). Four specimens were coated at each thickness. The system loss factor and natural frequencies were measured for each specimen to determine the material properties of the ceramic coating. This information was used with the dimensions at each stage to determine the material properties (storage modulus, loss modulus and loss factor) of the coatings. Differences in the results for different thickness specimens enable determining and quantifying any thickness effect.

BACKGROUND

In turbine engine applications, high cycle fatigue occurs when a component undergoes a level of dynamic stress, below the yield strength, that would cause failure in a number of cycles greater than or equal to 10^6 cycles. Damping can reduce the level of dynamic stress and therefore reduce the occurrences of high cycle fatigue failures. This is especially true when the engine speed causes an excitation at the resonant frequency of an engine component, leading to large deflections and stresses in the component. When the normal operating speed of the engine corresponds to the resonance of the blade, the life of the blade is limited. Possible methods to increase damping of vibrating systems include frictional dampers, constrained viscoelastic layers, air film dampers and ceramic hard coatings.

Ceramic hard coatings are typically composed of a metallic bond coat layer and an insulating ceramic coating layer. Benefits of these types of coatings are the resistance to heat transfer, corrosion, and erosion[1]. These coatings may be applied by one of several methods including air plasma spray and electron beam physical vapor deposition (EB-PVD)[2]. Material properties of coatings applied by air plasma spray and physical vapor deposition have been found to differ considerable due to the differences in the coating structures[3]. The storage modulus was found to be much higher and the loss modulus much lower when yttria stabilized zirconia was applied via EB-PVD as compared to air plasma spray. It has been shown the coatings made of plasma-sprayed materials can have high damping capacity[4]. The structure of such materials deposited from the plasma spray gun as droplets are believed to be made up of several small "splats" which can together act as multiple frictional dampers[5].

No data on the temperature dependence of the specific air plasma sprayed ceramic material used in this research (Titania-Alumina) is available. The temperature dependence of the damping capacity of similar materials has been considered by Patsias[6], who found the damping of a plasma sprayed hard coating to be constant with temperature to at least $400°C$. Tests conducted[7] with plasma sprayed Yttria Stabilized Zirconia, Magnesium Aluminate

Spinel, and Alumina showed no significant variation with temperatures to $300° F$ and strains as high as 1000 $\mu\epsilon$. Material Properties obtained for plasma sprayed magnesium aluminate spinel in three modes, covering frequencies from 300-1700 Hz, showed no significant difference[8].

One potential complication of ceramic hard coatings is that the application of relatively thick coatings will have a negative impact on the aerodynamics of the blade. To combat this problem it is possible to use design tools and engineer the coating to be applied in areas where damping is required most and with varying thickness. A very thorough understanding of the material properties of the coating and their variations as a function of thickness is essential to the success of the hard coating as a damping method in turbine engines. Previous studies by Patsias have suggested the thickness effect in a hard coating to be a volume effect[6]. Properties were determined on magnesium aluminate spinel of various thicknesses by Reed[9].

Methodologies have been developed for extracting material properties by comparing the response of a vibrating beam with a coating to the response of the same beam before the coating was applied. The classic method for extracting material properties is to assume that the system is described by the Öberst equations. These equations are only applicable for linear materials in which the loss modulus and storage modulus are independent of amplitude [10]. A method to extract material loss factors of coatings of more complex geometries by the use of modal strain energy[11] is also limited to linear materials. Another method considered, developed by Ustinov, involves intensive curve fitting and does not include the storage modulus of the material[12]. Patsias determined material properties using a partially covered beam[13]. A benefit of this method is that a partial coverage in a region of small strain gradient can be assumed to be in uniform strain. One potential issue with this method is that the effect of a partial patch on the mode shape of the coated beams does not easily allow the use of beam theory. Another issue is that with a small patch of coating the thicknesses would have to be large to see significant differences in the system response to determine properties.

The method chosen considers the difference in the system response of beams coated along both sides of the free length. This method considers the strain distribution in the beam, but assumes the mode shape for the coated and uncoated beams are the same. This method has been published elsewhere[14;15].

MATERIAL PROPERTIES

The mechanical properties determined in this analysis are the Young's storage modulus, loss modulus, and the material loss factor. This information is necessary in order to predict the how the coating will effect the behavior of any surface. With an understanding of the mechanical properties of damping coatings, it becomes possible for the manufacturer of an air foil to incorporate the behavior of the coating into the design.

The material properties will be defined as the storage modulus ,E_1, the loss modulus, E_2, and the material loss factor, η. The three are related through

$$E_2 = \eta E_1 \tag{1}$$

Here the storage modulus is the real part and the loss modulus is the imaginary part of a complex modulus which is defined in[10].

$$E^*(\epsilon, T) = E_1(\epsilon, T)[1 + j\eta(\epsilon, T)] = E_1(\epsilon, T) + jE_2(\epsilon, T) \tag{2}$$

EXPERIMENTAL RESEARCH

This research was conducted with the use of the equipment in the Turbine Engine Fatigue Facility (TEFF) at Wright Patterson Air Force Base, Dayton, Ohio. All of the testing was conducted with the use of the Unholtz-Dickie model SA15-560 6000-lb. shaker. The shaker was controlled using Vibration View software and uses a feedback control system which references an accelerometer at the base of the specimen. The shaker is capable of producing 6000 lbs. of force and has a frequency range of up to 3,000 Hz. The response of the specimens were measured using a Polytec Laser Vibrometer. The excitation frequency was slowly reduced from above resonance with excitation amplitude held constant. The resonant frequency (max in the response) was determined from the frequency response function. The system loss factor, η_{sys} as determined by the half-power bandwidth of the response function was used as the measure of damping. For a nearly linear single degree of freedom system, this can be related to the system quality factor through $\eta_{sys} = 1/Q$. The system loss factor is defined in equation 3.[16]

$$\eta_{sys} = \frac{D}{(2\pi U)} \tag{3}$$

Specimens

The specimens are nominally 25.4 cm long, 1.9 cm wide and 0.236 cm thick, but were measured individually for calculation of material properties. Two inches of the specimen are clamped leaving 20.32 cm of free length on the specimen. The substrate material, Ti 6 Al-4 V, is commonly used as in turbine engines, especially in stages of the low pressure compressors, the target application of the coating used for this research. Published[17] material properties for the substrate material are a storage modulus of 113.1 GPa and a density is 4.43 gm/cc.

The standard ceramic coating system includes the application of a bond coat to increase the adhesion of the coating to the beam. For this coating system the bond coat is air plasma sprayed NiCrAlY. This is a mixture of 22% Chromium, 10% Aluminum, 1% Yttrium. The balance is Nickel. Typically, the bond coat thickness is nominally 2-3 mils. The topcoat is a Titania-Alumina ceramic blend. The thicknesses used for the research were nominally 5 mils, 10 mils and 15 mils. The topcoat also is applied via an air plasma spray process.

Bare Beams

The natural frequencies of the bare specimens were measured and found to be within 3% of the theoretical values and to vary by less than 0.5%. This ensures that the boundary condition is sufficient for this type of testing.

The dependence on amplitude of the measured values of the loss factor, Figure 1, indicate system non-linearity. The damping doubles as the strain is increased from about 100 to about 500. This is primarily due to air damping[9]. The highest loss factor is 0.00064. In order for the error in the measurement of Q not to impact the results of the determination of material properties, the coated specimens must have significantly higher levels of damping. As it was expected that the loss factor for the coated beam would range from 0.005 to 0.01, the damping was considered low enough to enable determination of coating material properties.

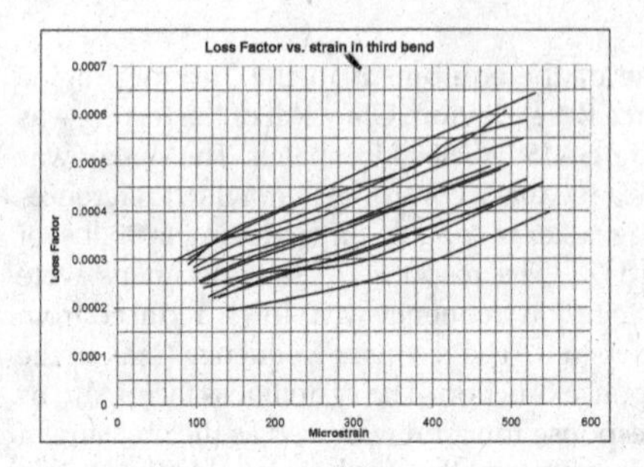

Figure 1: Loss factor for mode 3 in beams with no coating

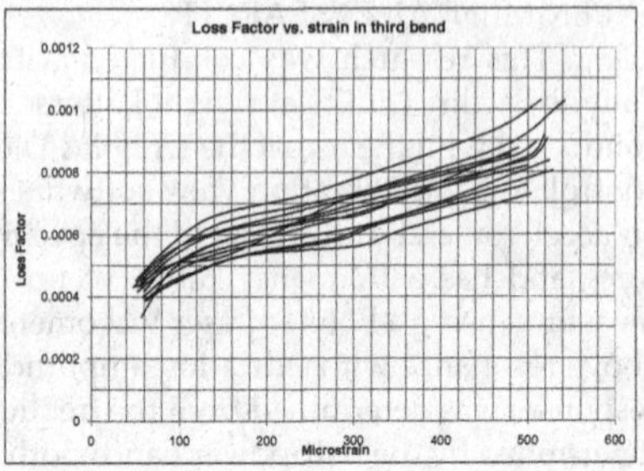

Figure 2: Loss factor for mode 3 in beams with bond coat only

Beams with Bond coat

All of the specimens were coated with approximately 3 mils of NiCrAlY bond coat on each side. There was a slight frequency increase in the specimens after the bond coat was applied. In the third mode, the frequency increased by about 3.3 Hz or an increase of 0.4%. This suggests that a small amount of stiffness was added to the system. The damping of the specimens with bond coat only is shown in Figure 2. The results show the highest loss factors to be around 0.001, or, at most only about twice the loss factor of the bare beam. In general, the bond coat does not significantly contribute to the damping of the coating system. The relatively small increases due to the bond coat damping precludes a valid extraction of the properties of the bond coat alone.

Beams with Bond Coat and Topcoat

The beams were then divided into three groups of four beams each and were coated with the Titania-Alumina blend ceramic coating. The beams were coated via an air plasma spray process to three different thicknesses by APS Materials, Inc. The nominal thicknesses chosen were 5, 10, and 15 mils. As 10 mils was the thickness most frequently used in prior testing, the influence of variations about that value are of special interest.

Very different frequencies result from the 3 coating thicknesses. The beam frequencies can be seen in Figure 3. For the 5, 10, and 15 mil specimens the average frequency reductions with increased strain were 1.75%, 2.7%, and 3.6%, respectively. As bare beam frequencies were between 795 and 815 Hz, and those for beams with bond coat between 797 and 818, it is evident that addition of the ceramic topcoat added significant stiffness to the system.

A transition from nonlinear (amplitude dependent) to nearly linear (amplitude independent) may be seen in the results in Figure 4. While each of the specimens shows a loss factor near 0.002 at low values of strain, for each of the thickness levels the loss factors asymptotically approach different values. The shape of the curve, an increase in damping at low strains and constant or diminishing values at higher strains, is typical of a ceramic hard coatings[3;8;9;14;15;18]. The specific irregularities in the data can be seen as a result of scatter in the data points. The results in mode 3 with the 10 mil coatings generally showed greater regularity.

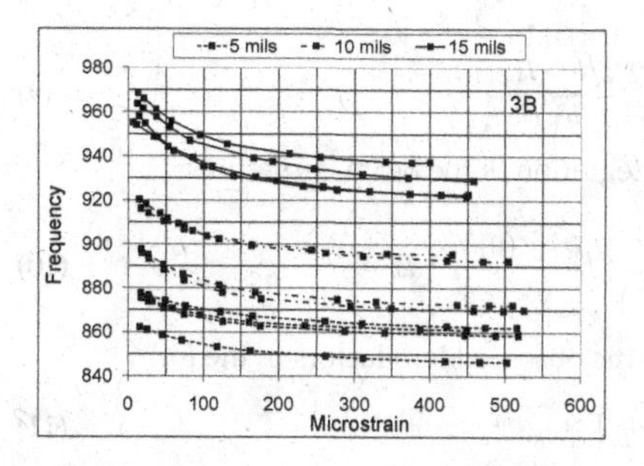

Figure 3: Frequency for mode 3 in beams with 5, 10 and 15 mils of top coat

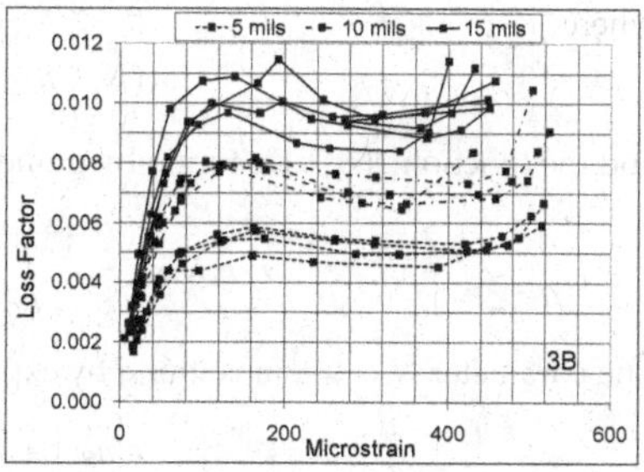

Figure 4: Loss factor for mode 3 in beams with 5, 10 and 15 mils of top coat

THE METHOD FOR DETERMINING MATERIAL PROPERTIES

The coating storage modulus can be found by comparing the resonant frequencies of the coated and uncoated beams. For a beam coated at thickness t on both sides of substrate thickness h, the ratio of energies stored in the coating and substrate is[15]

$$R_{SE} = \frac{U_{coating}}{U_{bare}} = \frac{f^2}{f_0^2}\left(1 + \frac{2t\rho_c}{h\rho_b}\right) - 1 \tag{4}$$

This leads to an average, or effective, value for the coating storage modulus, defined as

$$E_c = R_{SE}\frac{h}{6tT(2,t/h)}E_b \tag{5}$$

where E_b is the storage modulus of the bare beam and $T(2,t/h)$ is the evaluation for $N = 2$ of

$$T(N,t/h) = \frac{1}{t}\int_{h/2}^{h/2+t}\left(\frac{2z}{h}\right)^N dz = \frac{h}{2t}\frac{1}{N+1}[(1+2t/h)^{N+1} - 1] \tag{6}$$

For a non-linear coating material, the average, or effective, coating modulus, E_c, will vary with strain. This value is appropriate for use with similar strain distributions. A methodology for extracting the true material property from this average value has been developed[14].

The energy dissipated in the coating can be found by comparing the loss factors of the coated (η_s) and the uncoated (η_{bare}) beam at the same value of maximum interface strain at the root of the beam, ϵ_0. This can be used with the definition of the loss modulus in terms of the unit material damping (energy dissipated per unit volume per cycle)[16]

$$D = \pi E_2(\epsilon)\epsilon^2 \tag{7}$$

to obtain[15]

$$E_2 = \frac{D}{\pi\epsilon^2} = \frac{h}{6tT(2,t/h)}[M(N,t/h,n)][(1+R_{SE})\eta_S(\epsilon_0) - \eta_{BARE}(\epsilon_0)]E_{BARE} \tag{8}$$

where

$$M(N,t/h,n) \equiv \frac{T(2,t/h)}{T(N,t/h)}\frac{I(2,n)}{I(N,n)} \tag{9}$$

and the function $I(N,n)$ is found by numerical integration of the beam curvatures

$$I(N,n) = \frac{1}{L}\int_0^L \left[\frac{\partial^2 \chi_n(x)}{\partial x^2} / \frac{\partial^2 \chi_n(0)}{\partial x^2}\right]^N dx \tag{10}$$

The parameter $N = 2 + m$ is found by expressing the observed loss factors in the form

$$[\eta_S(1 + R_{SE}) - \eta_B] \cong \eta_0 \epsilon^m \tag{11}$$

For the data used in this study the non-linearity factor, $M(N,t/h,n)$, varied from 1.1 to 1.3. However, it has been shown[19] that the loss factors of non-linear systems as determined by bandwidth measurement are overestimated by 10-20% for the values of m seen in this work. Thus, the two errors are largely self cancelling and a satisfactory answer obtained by ignoring both[14].

Values of the loss modulus extracted by this methodology presume that system loss factors may be well represented by Lazan's power law representation.

$$D = J_\epsilon \epsilon^N \tag{12}$$

This has been found to adequately describe the damping versus stress for a wide range of structural materials at stresses below about 80% of their respective fatigue limits[20]. Other functional forms have also been considered[14;12].

EXPERIMENTAL RESULTS: MECHANICAL PROPERTIES

Density Calculations

Weights and thicknesses were determined for each of the specimens at every step of the process. Due to the nature of air plasma spray coatings, the surface can be very rough, as seen in the Scanning Electron Microscope (SEM) micrograph in Figure 5. As thickness measurements were made with a micrometer, values obtained are maximum values rather than average.

Values as found, however,were surprisingly consistent, as may be inferred from the standard deviations (SD) shown in Table 1. Values shown are averages for the 12 beams with bare and bond coat and for the 4 beams with each of the 3 coating thicknesses.

Titania-Alumina Properties

The material properties of these coatings may be interpreted in two ways. First, the response of the beams with both topcoat and bond coat may be compared to that of the bare beams. The material properties obtained are then average or effective values for the entire coating system. Second, the response of the beams with both bond and top coat can be compared with that of the beam with bond coat alone. The material properties obtained are then that for the top coat alone.

With the Titania-Alumina topcoat including bond coat treated as a single homogeneous

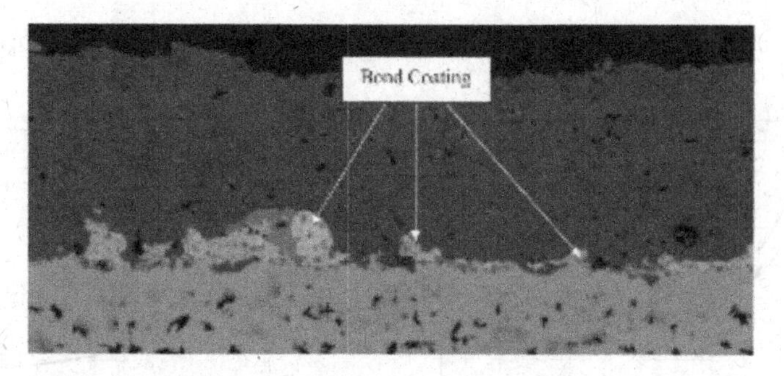

Figure 5: SEM of 10 mil coating (from Pearson[21])

Table 1: Error estimates related to density calculations

Component	N	Wt. (gm.)	SD(%)	Thick. (mils)	SD (%)	Dens. (gm/cc)	SD (%)
Bare	12	47.635	0.30%	90.6	0.31%	4.37	0.16%
Bond Coat	12	2.029	3.32%	3.7	2.6%	2.80	3.61%
5 mil	4	4.296	1.56%	5.4	1.32%	4.04	1.43%
10 mil	4	7.213	2.55%	9.6	1.65%	3.83	1.94%
15 mil	4	10.870	0.93%	14.9	1.41%	3.70	1.03%
System*	12	*for any thickness coating with bond coat				3.53	1.24%

system, all three thicknesses resulted in similar densities. The average density value for the homogeneous system is 3.53 gm/cc with a range of 3.84% and a standard deviation of 1.24%.

When the density of the Titania-Alumina topcoat only was calculated three different densities were found depending on the top coat thickness. The 5, 10 and 15 mil coatings yielded densities of 4.04, 3.83 and 3.7 gm/cc respectively, with ranges of 3.44%, 2.61 % and 1.1%. The differences appear to be statistically significant. But the apparent differences are a consequence of the difficulty in determining the true thickness of the bond coat. The typical SEM of Figure 5 suggests the average thickness may well have been about 1 mil greater than that obtained by subtracting the thicknesses with bond coat alone from the total thickness. A recalculation of densities accounting for the additional thickness gave values of 3.36, 3.48, and 3.47 gm/cc respectively. In the determination of material properties to follow, the actual dimensions and densities for the individual specimens were used.

Titania-Alumina Coating Including Bond Coat

The material properties for the coating as a whole (i.e. the equivalent homogeneous material) were first determined as this is most likely how the data would be used in prediction as the coating is used as a system. To do this, the coating storage modulus was found from Equation 5 with the coating thickness t taken as the combined thickness of bond and top coats, and the density as the effective value for the combination. The loss modulus was computed from Equation 8 using Equation 9 with the bare beam data used as the loss factor, η_{bare}.

The results showed that even though the storage modulus had a very strong strain dependence, all of the specimens showed similar values and trends over the range up to 500 $\mu\epsilon$ the storage modulus dropped about 12.5%.

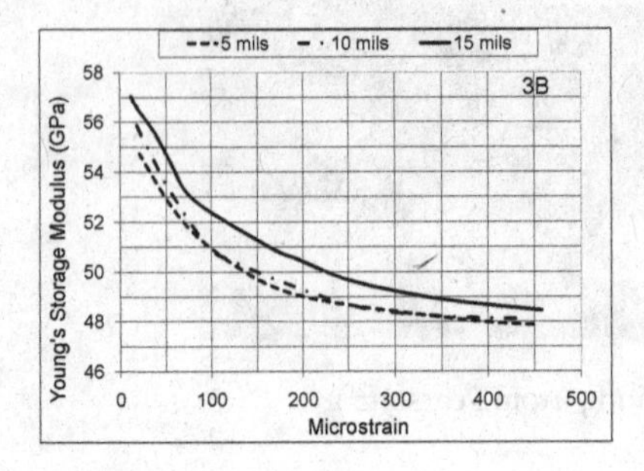

Figure 6: Titania-Alumina with bond coat storage modulus with 3 thicknesses

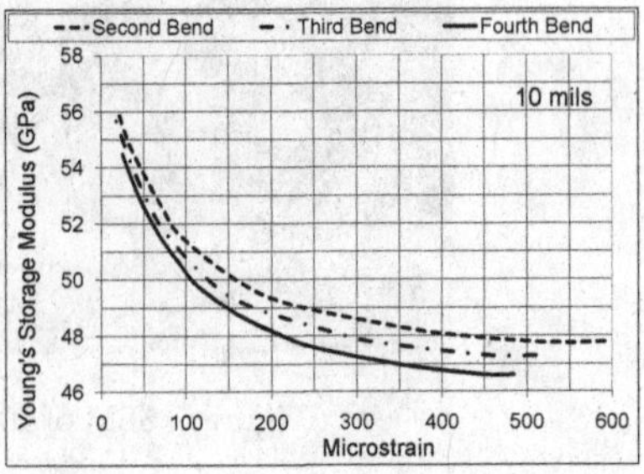

Figure 7: Titania-Alumina with bond coat storage modulus in three bending modes

Figure 6 was generated using one characteristic beam of each thickness. While the 15 mil coating does show a slightly higher storage modulus, the variation is minor. To determine if mode shape can affect the results, the results for one beam with a 10 mil coating in each of the three modes were considered (Figure 7). While the higher modes seem to show a lower storage modulus, the largest variation is only 3.2% and is not considered significant.

The loss modulus showed a similar trend. Each of the curves showed a rapid change until a strain near 75 $\mu\epsilon$, and then asymptotically approached a value near 1.03 GPa. A dependence of damping on the prior load history has been observed in ceramics[22;23]. Also, Reed[9] found the properties (modulus and loss factor) of a ceramic to vary over a history of several million cycles at high strain. To mitigate somewhat the influence of such history effects, the tests reported here were conducted with decreasing amplitudes of maximum strain. This test protocal is based on the assumption that the material is relatively stable at some level of stress if it has previously been subjected to loading at a much higher stress. However, as the tests conducted at the highest several amplitudes were not preceded by exposure to significantly higher amplitudes, the responses observed at the higher levels of strain (possibly above microstrains of 300-400 units) are not considered representative as the material is not likely to be stable. For this reason, the material properties presented here should not be regarded as reliable for strains above 300-400 $\mu\epsilon$. The irregularities in the curves are a result of the orignal data in Figure 4 having not been smoothed. The difference between the curves are not considered to be significant. However, mode 3 appeared to give superior data.

Titania-Alumina Coating Only

In this determination, the data taken from the tests with the bond coat only served as the bare beam for the proposes of extracting material properties. With this change, Equations 5 and 8 were applied as before.

When only the top coat is considered the storage modulus of the material was found to be slightly higher than that found for the equivalent homogeneous material. The apparent

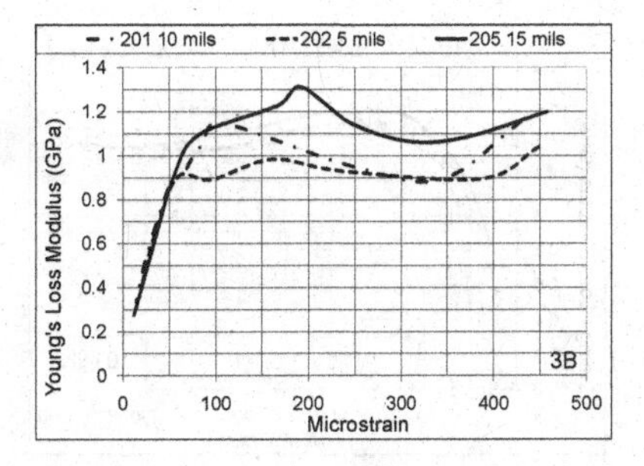

Figure 8: Titania-Alumina with bond coat loss modulus in third bend without the correction factor for three coating thicknesses

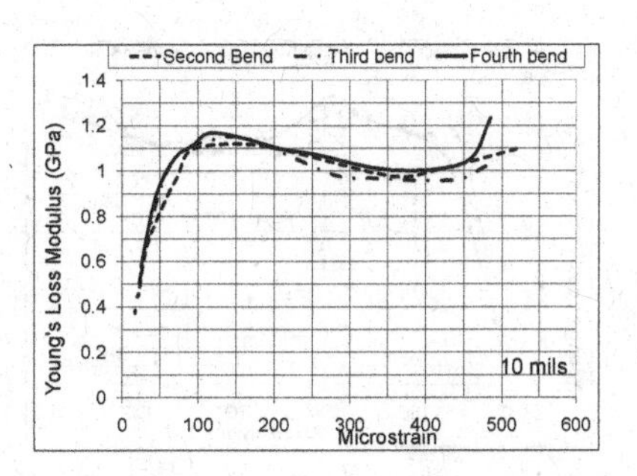

Figure 9: Titania-Alumina with bond coat loss modulus in three bending modes without the correction factor

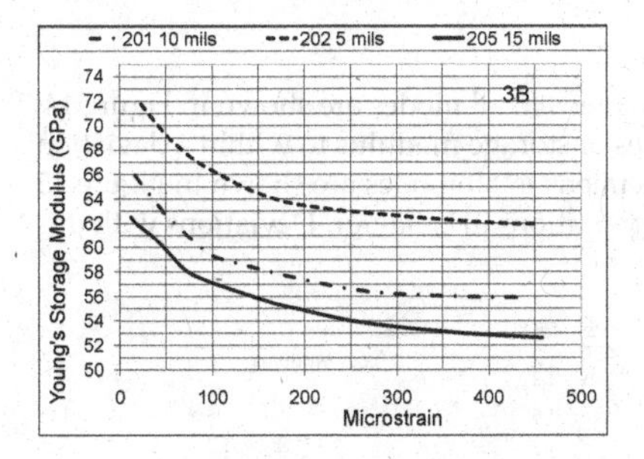

Figure 10: Titania-Alumina only storage modulus with three thicknesses

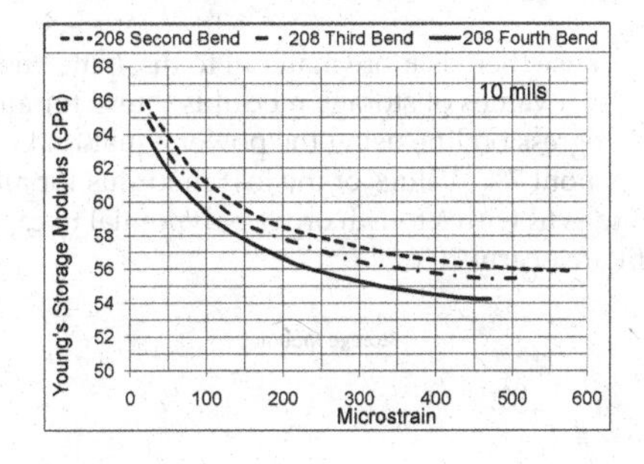

Figure 11: Titania-Alumina only storage modulus in three modes

values are given in Figure 10.

At 60 $\mu\epsilon$, the values for storage modulus for the 5, 10 and 15 mil coatings were found to be approximately 68.95, 62.05 and 58.61 GPa respectively. It is interesting to note that if these values are scaled as were the densities to account for the estimated 1 mil of error in the measurement the values become 57.43, 56.4 and 55.16 GPa. These differences are not considered significant.

It can be seen in Figure 11 that for coatings of the same thickness the difference in the properties determined from data taken in the three modes are not seen as significant.

The loss modulus determined for all the thicknesses were similar. Results for typical specimens of each coating thickness are shown in Figure 12. Results for the same specimen in the three modes, Figure 13, are very similar.

RESULTS AND CONCLUSIONS

This research shows that the damping effect of Titania-Alumina ceramic coatings is a volume effect. For each thickness and mode, the calculated material properties remained unchanged. This shows that while the thicker coating did show higher damping this is simply

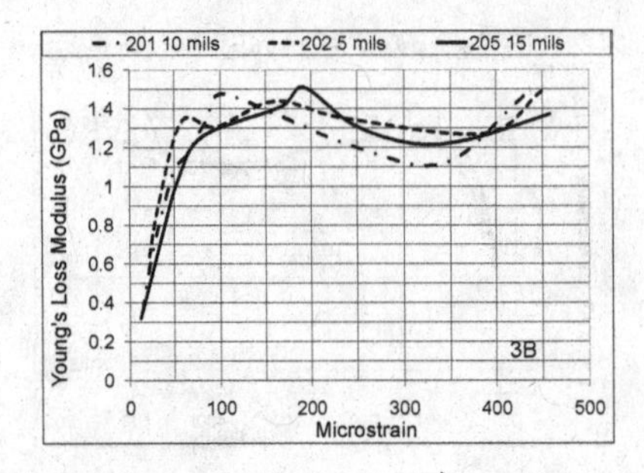

Figure 12: Titania-Alumina only loss modulus in third bend without the correction for three coating thicknesses

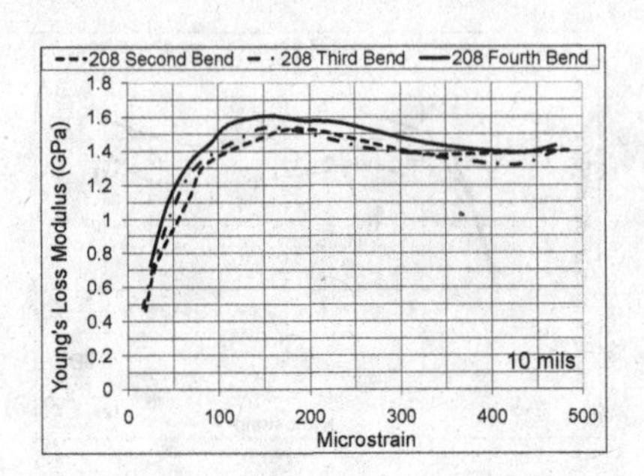

Figure 13: Titania-Alumina only loss modulus in three bending modes without the correction factor

because there is more material to dissipate energy.

Values of storage modulus found for all specimens in all modes are shown in Figure 14. A regression line using the power equation fit values of storage modulus to within a deviation of about 7%. Values of the loss modulus for all specimens in all modes are shown in Figure 15 and is fit with a fourth order polynomial to capture variations in behavior. Deviations with this fit are nearly 20%.

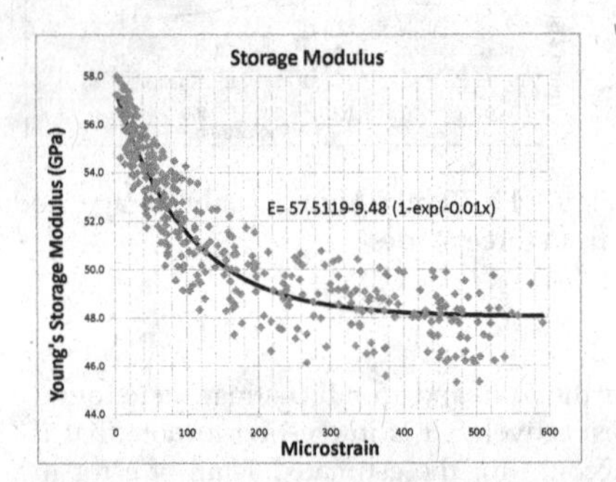

Figure 14: Titania-Alumina storage modulus for all beams in all bending modes

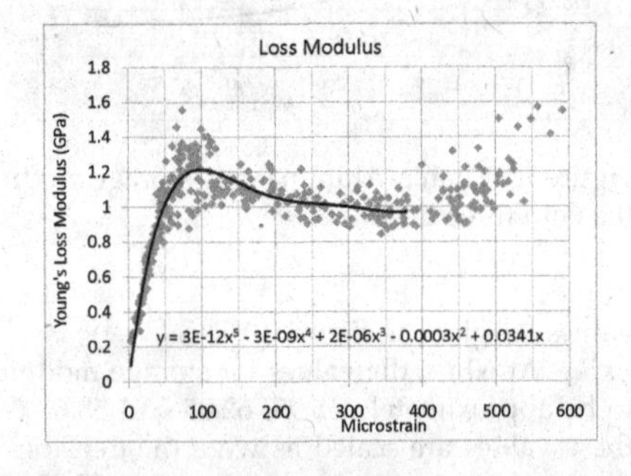

Figure 15: Titania-Alumina loss modulus for all beams in all bending modes

These results confirm the preliminary findings of Patsias[6], who suggested that damping was proportional to coating thickness.

BIBLIOGRAPHY

[1] T Strangman. Thermal barrier coatings for airfoils. Thin Solid Films, 127:93-105, 1985.

[2] B Movchan and A Ustinov. Highly damping hard coatings for protection of titanium blades. Presented at the RTO AVT-121 Symposium on Evaluation, Control and Prevention of High Cycle Fatigue in Gas Turbine Engines for Land, Sea and Air Vehicles held in Seville, Spain, pages 11-1 - 11-15, October 3-5 2005.

[3] N Tassini, K Lambrinou, I Mircea, Sophoclies Patsias, O. Van der Biest, and R Stanway. Comparison of the damping and stiffness properties of 8 wt percent yttria stabilized zirconia ceramit coat deposited by the aps and eb-pvd techniques. Smart Structures and Materials 205: Damping and Isolation, edited by Kon- Weil Wang, Proceeding of SPIE, 5760, SPIE:109-117.
[4] K Cross, W Lull, R Newman, and J Cavanagh, Potential of graded coatings in vibration damping. Journal of Aircraft, 10(11):689-691, November 1973.
[5] F Kroupa and J Plesek, Nonlinear elastic behavior in compression of thermally sprayed materials. Materials Science and Engineering, A328:1-7, 2002.
[6] S Patsias, G Tomlinson, and M Jones. Initial studies into hard coating for fan blade damping. In Proceedings of the 6th National Turbine Engine High Cycle Fatigue (HCF) Conference, Jacksonville, FL, March 5-8, 2001.
[7] M Willson, R Willson, J Henderson, A Nashif, P Torvik, and D Zabierek. Damping Coatings for Gas Turbine Compression System Airfoils: Final Report for SBIR phase I project, APS Materials Inc., contract no. :N68335-04-C-0129, Topic No.: NO4-019, (version approved for public release). September 6, 2004.
[8] P Torvik. Evaluation of Damping Properties of Coatings, Part I: Room temperature. Universal Technology Corporation, March 2002.
[9] S Reed. Development of Experimental, Analytical, and Numerical Approximations Appropriate for Nonlinear Damping Coatings. PhD thesis, Air Force Institute of Technology, Wright-Patterson Air Force Base, Ohio, November 2007.
[10] A Nashif, D Jones, and J Henderson. Vibration Damping. John Wiley and Sons, 1985.
[11] C Johnson and D Keinholz, Finite element predictions of damping in structures with constrained viscoelastic layers. AIAA Journal, 20(9):1284-1290, September 1982.
[12] A Ustinov, V Skorodzievskii, and N Kosenko. A Study of the Dissipative Properties of Homogeneous Materials Deposited as Coatings. Method for the Determination of the Amplitude Dependence of the True Vibration Decrement of the Coating Material. Strength of Materials, 39(6):663-670, 2007.
[13] S Patsias, C Saxton, and M Shipton. Hard damping coatings: An Experimental Procedure for Extraction of Damping Characteristics and Modulus of Elasticity. Materials Science and Engineering A, 370:412-416, 2004.
[14] P Torvik. Determination of Mechanical Properties of Non-Linear Coatings from Measurements with Coated Beams. International Journal of Solids and Structures, Vol. 46, No. 5 (1066-1077) March 2009.
[15] P Torvik, R Willson, and J Hansel. Influence of a viscoelastic surface inltrate on the damping properties of plasma sprayed alumina coatings part I: Room temperature. In Proceedings, Materials Science and Technology 2007 Conference and Exhibition, (MST 2007), Sept 15-20, Detroit, MI.
[16] R Plunkett. Measurement of Damping (117-132). ASME New York, 1959.
[17] Department of Defense Handbook: Metallic Materials and Elements for Aerospace Vehicle Structures, volume MIL-HDBK-5J. Department of Defense, 31 January 2003.
[18] S Patsias, N Tassini, and R Stanley. Hard ceramic coatings: An experimental study on a novel damping treatment. In Kon-Well Wang, editor, Proceedings of SPIE: Smart Structures in materials 2004: Damping and Isolation, volume 5386, pages 174-184. SPIE, 2004.
[19] P Torvik. A note on the estimation of non-linear system damping. Journal of Applied Mechanics, ASME, 70, May 2003.
[20] B Lazan. Damping of Materials and Members in Structural Mechanics. Pergamon Press, Oxford, 1968.

[21] L Pearson. Vibration analysis of Commercial Thermal Barrier Coatings. Master's thesis, Air Force Institute of Technology, June 2008.
[22] Tassini, N., Lambrinou, K., Mircea, I., Bartsch, M., Patsias, S., and Van der Biest, O., Study of the Amplitude-Dependent Mechanical Behaviour of Yttria-stabilised Zirconia Thermal Barrier Coatings. Journal of the European Ceramic Society, 27(2-3), pp. 1487-1491. 2007
[23] Patsias, S., and Williams, R. J.,Hard damping coatings: Material Properties and F.E. Prediction Methods. Proceedings of the 8th HCF COnference, Monterey, CA, pp. 5d1-17. 2003

ELECTRICAL AND DIELECTRIC PROPERTIES OF THERMALLY GROWN OXIDE (TGO) ON FECRALLOY SUBSTRATE STUDIED BY IMPEDANCE SPECTROSCOPY

Fan Yang, Akio Shinmi and Ping Xiao*
Materials Science Centre, School of Materials, University of Manchester
Manchester M1 7HS, UK

ABSTRACT

Thermally grown oxide (TGO) formed on Fecralloy substrate exposed to an oxidizing environment of 1100°C for 4 hours was characterized by impedance spectroscopy method. Electrical and dielectric properties of TGO were investigated. AC conductivities and dielectric constants were studied as a function of frequency and temperature. AC conductivity analysis indicates a change in conduction mechanism around 400°C. Dielectric constant shows remarkable dispersions at low frequencies and high temperatures, which is considered to be originated from the Maxwell-Wagner type polarization from the interface of grain and grain boundary.

INTRODUCTION

Thermal barrier coatings (TBCs) have been widely used in the hot section of aeroturbine engines to increase turbine efficiency and to extend the life of metallic components [1-3].TBCs have a multi-layered structure, typically consisting of a ceramic topcoat yttria stabilized zirconia (YSZ), a metallic bond coat, and a thermally grown oxide (TGO) layer formed at the topcoat/bond coat interface due to the oxidation of the bond coat at high temperature. The TGO, predominantly alumina, provides oxidation protection and has a major influence on the TBC durability [4]. Since TGO is a dominant factor leading to failure of TBCs [5], a deep understanding of the properties of the TGO is a critical issue in TBCs.

Impedance spectroscopy is a powerful method to characterize the electrical and dielectric properties of solid materials and their interfaces. Since 1999, impedance spectroscopy has been developed to examine degradation of TBCs as a non-destructive evaluation tool, which is critical for prediction of TBCs lifetime during service [6-16]. Much effort has been devoted to characterize the TGO layer in TBCs by impedance spectroscopy. For example, Song *et al.* studied the formation and growth of TGO layer in air plasma sprayed (APS) TBCs by impedance spectroscopy and found the formation, thickening, densification and composition change of the TGO layer during TGO growth can be qualitatively evaluated by the peak shifts on the Bode plot [11], and the results were confirmed by finite element simulation by Deng *et al.* [17]; Wang *et al.* also studied the TGO growth in TBCs by impedance spectroscopy and found a linear relationship between the TGO thickness and the diameter of the semicircle on the electrical modulus spectra, which proved the thickness of TGO layer in TBCs can be estimated by measuring its impedance response [18]. The above studies may provide useful information in estimating the remaining lifetime of the TBC system. However, it is still necessary to characterize the TGO itself to obtain a better and clear understanding of the TGO layer inside the TBCs.

Fecralloy is a high temperature oxidation-resistant alloy. Under high temperature, a continuous, adherent, slowly-growing alumina surface scale is formed on Fecralloy, which provides excellent oxidation resistance against a high temperature and a corrosive atmosphere [19]. A preliminary impedance spectroscopy study on thermally oxide alumina scale on Fecralloy was performed by Deng *et al.* [20]. It was found the change of impurity level in the TGO can be monitored by according to variation of activation energy. However, the conduction mechanisms of the TGO layer remains unclear.

Therefore in this paper, the electrical and dielectric properties of the TGO scale on Fecralloy substrate were studied using impedance spectroscopy. The TGO sample was formed by oxidizing the Fecralloy plate at 1100°C for 4 hours in air. Impedance measurements were carried out from 250 to 500°C over a frequency range from 0.1 to 10^6 Hz. AC conductivities and dielectric constants were studied as a function of frequency and temperature, which can provide useful information on the conduction or polarization mechanisms of the TGO. The aim of this study is to provide preliminary and fundamental understanding of electrical and dielectric properties of the thermally grown alumina, and thus contribute to further studying of TGO growth and non-destructive evaluation of thermal barrier coatings.

EXPERIMENTS

Commercial Fecralloy plate with a thickness of 1.0 mm (Goodfellow Metals Ltd., UK) was cut into 10 mm × 8 mm rectangles, which were mechanically polished up to 0.25 μm with diamond paste and then cleaned in acetone with ultrasonic vibration. Oxidation was carried out in air at 1100°C for 4 hours, with heating and cooling rates of 3°C/min and 30°C/min, respectively. The thickness of the TGO layer was around 700 nm, evaluated by the fractured cross-section observation under scanning electron microscope (SEM).

Electrical and dielectric properties of the TGO sample were obtained from AC impedance spectroscopy measurements using a Solatron SI 1255 HF frequency response analyser coupled with a Solatron 1296 Dielectric Interface (Solartron, UK). For impedance measurement, one side of the oxidized Fecralloy substrate was carefully polished away to expose the metal surface, which served as the bottom electrode. Ag paint was coated on the surface of the TGO and fired at 690°C for 30 mins as the top electrode. An AC voltage of 0.05 V was applied to the sample over a frequency range from 0.1 to 10^6 Hz at various temperatures from 250°C to 500°C. Equivalent circuit fittings of the measured impedance spectra were carried out using Zview Impedance Analysis software (Scribner Associates, Inc., Southern Pines, NC).

RESULTS AND DISCUSSION

Impedance and electrical modulus spectra

Figure 1 shows the Nyquist plots for the TGO sample measured at different temperatures. The features of the Nyquist plot change obviously with the increase of temperature. At low temperatures (below 400°C), there is only one distorted semicircle on the Nyquist plot, while at high temperatures, more arcs appear.

A typical Nyquist plot of a ceramic material includes three responses: one semicircle at high frequency represents the response from grains, one semicircle at medium frequency represents the response from grain boundaries and a tail or an arc at low frequency represents the electrode effect [21]. The Nyquist plots measured at 450 and 500°C in Figure 1(e) and (f) correspond well with the above characteristics and can be fitted by an equivalent circuit of three R (resistance) -CPE (constant phase element) elements in series connection. The black lines with triangle symbols in Figure 1(e) and (f) are the fitting curves, indicating the measured impedance spectra are well fitted by this equivalent circuit. However, when the temperature is below 400°C, it is difficult to distinguish the responses from different components only from Nyquist plots because of severe overlapping between each component. Therefore it is necessary to plot the electrical modulus spectra to make it clear how many relaxation processes are actually included.

Figure 2 shows the electrical modulus spectra for the TGO sample measured at different temperatures. Electrical modulus can be calculated from the complex impedance by the following equation:

$$M = M'+jM''= j\omega C_0 Z = j\omega C_0(Z'-jZ'') = \omega C_0 Z''+j\omega C_0 Z' \quad (1)$$

Here ω is the angular frequency, C_0 is the vacuum capacitance of the empty measuring cell, which can be calculated by $C_0 = \varepsilon_0 \frac{A}{d}$, where ε_0 is the vacuum permittivity (8.85×10^{-12} F/m), A is the electrode area and d is the thickness of the TGO layer. From Fig.2, it can be seen that there are two depressed semicircles overlapped to certain extent, indicating two relaxation processes are included. Therefore, the impedance spectra measured at low temperatures (at and below 400°C) can be fitted by an equivalent circuit of two R-CPE elements in series connection. The fitting results are shown in the black lines with triangle symbols in Figure 1(a)-(c), which fit the experimental results very well.

In combination of Figure 1(a)-(c), Figure 2 and equivalent circuit fitting, it is confirmed that the single semicircle on the Nyquist plot for the TGO sample measured below 400°C includes two relaxation processes. These two relaxation processes have similar time constants, leading to severe overlap on the Nyquist plot. When temperature increases to 400°C, the response from each relaxation process becomes distinguishable, and therefore leads to two well resolved semicircles. With further increase of temperature, the contribution of electrode effect occurs within the measuring frequency range, and as a result, another arc appears at low frequencies.

AC conductivity

In order to study the conduction mechanism in the TGO sample, AC conductivity analysis is performed. AC conductivity of the TGO can be obtained from the impedance data by the following relationship:

$$\sigma'_{AC} = \frac{d}{A} \cdot \frac{Z'}{(Z')^2 + (Z'')^2} \tag{2}$$

where d is the thickness of the TGO layer, and A is the electrode area.

Figure 3 shows the frequency and temperature dependence of the AC conductivity. It can be seen that the AC conductivity of the TGO sample varies from 10^{-11} to 10^{-5} S/cm over the temperature range from 250 to 500°C and frequency range from 0.1 to 10^6 Hz. The electrical conduction in the TGO sample is much more temperature dependent at low frequencies than that at high frequencies.

For a clear and detailed description, we extracted the AC conductivity of the sample measured at 350°C and plotted in the top-left inset, which shows a characteristic electrical response of ionic conductors [22, 23]. It can be seen that the variation of AC conductivity with frequency can be generally divided into three regions. At low frequencies, there is a plateau where the slope of $\log\sigma$ versus $\log f$ is zero, reflecting macroscopic charge transportation and giving the value of DC conductivity [24]. With the increase of frequency, the DC plateau crosses over into a dispersive conductivity region, where the relationship between the AC conductivity and frequency obeys the well known Jonscher power law, which is also termed "Universal Dynamic Response (UDR)", with the expression of $\sigma(f) = Af^n (0 < n < 1)$, where A is a temperature-dependent constant and n is a fractional exponent [22, 25]. On a double logarithmic plot, the conductivity has a linear relationship with frequency with a slope of n. Usually it is believed the UDR region reflects the hops of mobile ions [26] or the influence of ion-ion correlation on ion motion [22, 27, 28]. Further increase of frequency leads to a "Nearly Constant Loss (NCL)" region [29, 30], where the AC conductivity has a linear relationship with frequency with a slope of 1 in a double logarithmic plot. The NCL region reflects the localized motion of charge carriers [31, 32].

The bottom-right inset in Figure 3 shows the slope of the UDR region at various temperatures. It can be seen that, below 350°C, the slope of the UDR region declines with the increase of temperature. However, with further increase of temperature, the slope of the UDR region goes upward. It is considered that, the slope of this UDR region is influenced by two competitive factors. One is the motion of charge carriers, which is much easier at high temperatures and therefore leads to a decrease in the slope of UDR region with increasing temperature. On the other hand, however, with the increase of temperature, the correlation between the charge carriers becomes dominant, and therefore results in an increase of the slope. Consequently, the slope of the UDR region shows a downward followed by an upward trend with the increase of temperature.

In Figure 3, the AC conductivity of the TGO sample measured at 450 and 500°C shows a further decrease after the DC plateau at very low frequencies, which is possibly caused by the presence of blocking effects from grain boundaries or electrodes [21]. The DC conductivity of the TGO sample is obtained from the plateau in the AC conductivity plot, and its relationship with the reciprocal of temperature is shown in Figure 4. The conductivity at low temperatures (250, 300 and 350°C) has a linear relationship with the reciprocal of absolute temperature and gives an activation energy of 0.82eV. The conductivity at high temperatures (400, 450 and 500°C) also obeys the Arrhenius law, and gives an activation energy value of 1.77eV. The change of activation energy within the temperature range can be a result of either a change of conduction mechanism or a change of charge carriers. The turning point is around 400°C.

It is generally accepted that the electrical conduction in alumina follows the trend: ionic conduction ($T < 400°C$), mixed ionic and electronic conduction ($400 < T < 800$-$1000°C$), electronic conduction ($T > 800$-$1000°C$). Therefore it is reasonable to consider the turning point in Fig.4 indicates a change of conduction mechanism from ionic to mixed conduction. It should be noted that, the conductivity of the TGO sample is several orders higher than the reported values of pure alumina. For example, Will *et al.* studied the electrical conduction mechanism in high-purity single-crystal alumina within a wide temperature range from 400 to 1300°C under a DC electric field [33]. In their study, the magnitude of electrical conductivity at 400°C is as low as 10^{-16} S/cm, and it gives activation energy value of 0.4 – 0.9 eV within the temperature range from 400 to 700°C. Other researchers [34] also reported the electrical conductivity in single crystal alumina, which is 4 orders higher than the value given by Will *et al.*, but still 3 orders lower than our result. One possible reason for this high conductivity is the impurities inside the TGO. Our previous study [20] found Fe, Cr, Ti or Zr impurities inside the thermally oxidized alumina scale on Fecralloy substrate. These impurities can generate acceptor or donor level in the band gap of alumina, and therefore enhance the electronic conduction at high temperatures. It is also possible that the existence of impurities increases the vacancy concentration of the TGO and increase the ionic conductivity at low temperatures. However, due to the complicated microstructure and composition of the TGO, further effort is still needed for clarifying the origin of the high conductivity.

Dielectric properties

The dielectric constant (real part of the complex dielectric constant) can be obtained from the impedance data by the following relationship:

$$\varepsilon' = \frac{1}{\omega C_0} \frac{Z''}{(Z')^2 + (Z'')^2} = \frac{d}{A\omega\varepsilon_0} \frac{Z''}{(Z')^2 + (Z'')^2} \quad (3)$$

where ω is the angular frequency, d represents the thickness of the TGO layer, A represents the electrode area, and ε_0 is the vacuum dielectric constant (8.85×10^{-12} F/m).

The frequency dependent dielectric constant of the TGO sample at different temperatures is shown in Figure 5. It can be seen that the dielectric constant exhibits remarkable dielectric dispersion, especially at high temperatures and low frequencies. In a dielectric material, four different types of polarization mechanisms are possible: electronic and ionic relaxations, which are related to rapid oscillations of weak dipoles and occur at very high frequencies ($>10^{10}$ Hz); dipolar polarization that appears in the frequency range from 100 to 10^7 Hz; space charge or interfacial polarization occurring at low frequencies and high temperature, which are related to ionic space charge carriers, interface polarization located at grain boundary or the interface between the sample and electrode, and electrode polarization [35]. In case of TGO, it is considered the ultrahigh dielectric constant at low frequencies and high temperatures originates from the typical Maxwell-Wagner type polarization from the interface of grain and grain boundary rather than the electrode polarization, since at 350°C, when the electrode effect is eligible, an obvious increase in dielectric constant can still be observed in Figure 5 at low frequency.

CONCLUSION

Electrical and dielectric properties of thermally grown oxide on Fecralloy substrate have been studied by impedance spectroscopy method. AC conductivities and dielectric constants were studied as a function of frequency and temperature. Complex impedance and electrical modulus spectra show two relaxation processes when measuring temperature is below 400°C and three responses at higher temperatures. AC conductivity analysis indicates a change in conduction mechanism around 400°C. Dielectric constant shows remarkable dispersions at low frequencies and high temperatures, which is considered to be originated from the Maxwell-Wagner type polarization from the interface of grain and grain boundary.

REFERENCES

[1]R. A. Miller, "Current Status of Thermal Barrier Coatings – An Overview", *Surf. Coat. Tech.*, **30,** 1-11 (1987)

[2]N. P. Padture, M. Gell and E. H. Jordan, "Thermal Barrier Coatings for Gas-Turbine Engine Applications", *Science*, **296,** 280-84 (2002)

[3] D. R. Clarke and C. G. Levi, "Materials Design for the Next Generation Thermal Barrier Coatings", *Annu. Rev. Mater. Res.*, **33,** 383-417 (2003)

[4] S. Lee, S. Sun and K. Kang, "In-situ Measurement of the Thickness of Aluminium Oxide Scales at High Temperature", *Oxid. Met.*, **63,** 73-85 (2005)

[5] A. G. Evans, D. R. Mumm, J. W. Hutchinson, G. H. Meier and F. S. Pettit, "Mechanisms Controlling the Durability of Thermal Barrier Coatings", *Prog. Mater. Sci.*, **46,** 505-553 (2001)

[6] K. Ogawa, D. Minkov, T. Shoji, M. Sato and H. Hashimoto, "NDE of Degradation of Thermal Barrier Coating by Means of Impedance Spectroscopy", *NDT&E Int.*, **32,** 177-85 (1999)

[7] B. Jayaraj, V. H. Desai, C. K. Lee and Y. H. Sohn, "Electrochemical impedance spectroscopy of porous ZrO2-8wt.% Y2O3 and thermally grown oxide on nickel aluminide", *Mater. Sci. Eng. A*, **372,** 278-286 (2004)

[8] B. Jayaraj, S. Vishweswaraiah, V. H. Desai and Y. H. Sohn, "Electrochemicala impedance spectroscopy of thermal barrier coatings as a function of isothermal and cyclic thermal exposure", *Surf. Coat. Tech.*, **177-178,** 140-151 (2004)

[9] N. Q. Wu, K. Ogawa, M. Chyu and S. Mao, "Failure Detection of Thermal Barrier Coatings Using Impedance Spectroscopy", *Thin Solid Films*, **257,** 301-06 (2004)

[10] M. S. Ali, S. Song and P. Xiao, "Evaluation of Degradation of Thermal Barrier Coatings Using Impedance Spectroscopy", *J. Eur. Ceram. Soc.*, **22,** 101-07 (2002)

[11] S. Song and P. Xiao, "An Impedance Spectroscopy Study of High-temperature Oxidation of Thermal Barrier Coatings", *Mater. Sci. Eng.B*, **97,** 46-53 (2003)

[12] X. Wang, J. Mei and P. Xiao, "Determining Oxide Growth in Thermal Barrier Coatings (TBCs) Non-destructively Using Impedance Spectroscopy", *J. Mater. Sci. Lett.*, **20,** 47-49 (2001)

[13] J. W. Byeon, B. Jayaraj, S. Vishweswaraiah, S. Rhee, V. H. Desai and Y. H. Sohn, "Non-destructive Evaluation of Degradation in Multi-layered Thermal Barrier Coatings by Electrochemical Impedance Spectroscopy", *Mater. Sci. Eng. A*, **407,** 213-25 (2005)

[14] Y. H. Sohn, B. Jayaraj, S. Laxman, B. Franke, J. W. Byeon and A. M. Karlsson, "The Non-Destructive and Nano-Microstrutural Characterization of Thermal-Barrier Coatings", *JOM*, **56,** 53-56 (2004)

[15] B. Jayaraj. S. Vishweswariah, V. H. Desai and Y. H. Sohn, "Changes in Electrochemical Impedance with Microstructural Development in TBCs", *JOM*, **58,** 60-63 (2006)

[16] J. Zhang and V. Desai, "Evaluation of Thickness, Porosity and Pore Shape of Plasma Sprayed TBC by Electrochemical Impedance Spectroscopy", *Surf. Coat. Tech.*, **190,** 98-109 (2005)
[17] L. Deng, Y. Xiong and P. Xiao, "Modelling and Experimental Study of Impedance Spectra of Electron Beam Physical Vapour Deposition Thermal Barrier Coatings", *Surf. Coat. Tech.*, **201,** 7755-63 (2007)
[18] X. Wang, J. Mei and P. Xiao, "Non-destructive Evaluation of Thermal Barrier Coatings Using Impedance Spectroscopy",*J. Eur. Ceram. Soc.*, **21,** 855-59 (2001)
[19] W. J. Quadakkers, "Growth mechanisms of oxide scales on ODS alloys in the temperature range 1000-1100°C", *Mater. Corr.*, **41,** 659-68 (1990)
[20] L. Deng and P. Xiao, "Characterisation of alumina scales on Fecralloy using impedance spectroscopy", *Thin Solid Films*, **516**, 5027-31 (2008)
[21] J. R. MacDonald, Impedance Spectroscopy – Emphasizing Solid Materials and Systems. John Wiley & Sons, New York, USA, 1987
[22] A. K. Jonscher, "The 'Universal' Dielectric Response", *Nature*, **267,** 673-79 (1977)
[23] J. Garcia-Barriocanal, A. Rivera-Calzada, M. Varela, Z. Sefrioui, E. Iborra, C. Leon, S. J. Pennycook and J. Santamaria, "Colossal Ionic Conductivity at Interfaces of Epitaxial ZrO_2: Y_2O_3/$SrTiO_3$ Heterostructures", *Science*, **321,** 676-80 (2008)
[24] K. Funke and R. D. Banhatti, "Translational and Localised Ionic Motion in Materials with Disordered Structures", *Solid State Sci.*, **10,** 790-803 (2008)
[25] A. K. Jonscher, "The Interpretation of Non-Ideal Dielectric Admittance and Impdance Diagrams", *Phys. Status Solidi A*, **32,** 665-76 (1975)
[26] D. P. Almond, G. K. Duncan, A. R. West, "The Determination of Hopping Rates and Carrier Concentrations in Ionic Conductors by a New Analysis of ac Conductivity", *Solid State Ionics*, **8,** 159-64 (1983)
[27] K. L. Ngai and A. K. Rizos, "Parameterless Explanation of the Non-Arrhenius Conductivity in Glassy Fast Ionic Conductors", *Phys. Rev. Lett*, **76,** 1296-99 (1996)
[28] K. L. Ngai, G. N. Greaves and C. T. Moynihan, "Correlation between the Activation Energy for Ionic Conductivity for Short and Long Time Scales and the Kohlrausch Stretching Parameter β for Ionically Conducting Solids and Melts", *Phys. Rev. Lett*, **80,** 1018-21 (1998)
[29] W. K. Lee, J. F. Liu and A. S. Nowick, "Limiting Behavior of ac Conductivity in Ionically Conducting Crystals and Glasses: A New Universality", *Phys. Rev. Lett*, **67,** 1559-61 (1991)
[30] K. L. Ngai, "Properties of the Constant Loss in Ionically Conducting Glasses, Melts, and Crystals", *J. Chem. Phys.*, **110,** 10576 (1999)
[31] M. Pollak and G. E. Pike, "ac Conductivity of Glasses", *Phys. Rev. Lett*, **28,** 1449-1451 (1972)
[32] A. S. Nowick, "Exploring the Low-temperature Electrical Relaxation of Crystalline Oxygen-ion and Protonic Conductors", *Solid State Ionics*, **136-137,** 1307-14 (2000)
[33] F. G. Will, H. G. deLorenzi and K. H. Janora, "Conduction Mechanism of Single-Crystal Alumina", *J. Am. Ceram. Soc*, **75,** 295-304 (1992)
[34] H. M. Kizilyalli and P. R. Mason, "D.C. and A.C. Electrical Conduction in Single Crystal Alumina", *Phys. Status Solidi A*, **36,** 499-508 (1976)
[35] S. Lee, K. Kang and S. Han, "Low-frequency Dielectric Relaxation of $BaTiO_3$ Thin-film Capacitors", *Appl. Phys. Lett*, **75,** 1784-86 (1999)

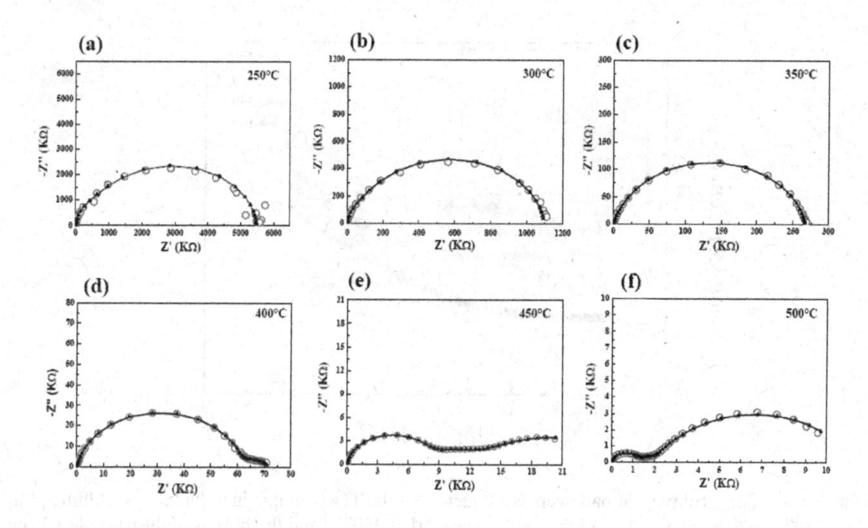

Figure 1. Impedance spectra (Nyquist plot) for the TGO sample measured at various temperatures. (a) 250°C (b) 300°C (c) 350°C (d) 400°C (e) 450°C (f) 500°C. The hollow circles represent for measured data; the black lines with triangles are equivalent fitting results.

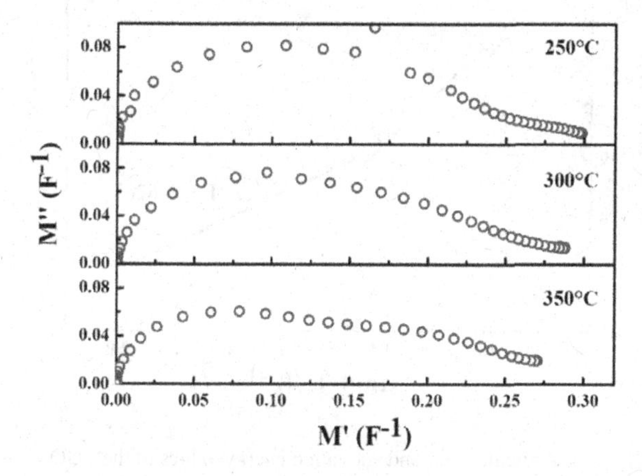

Figure 2. Modulus spectra of the TGO sample measured at 250, 300 and 350°C, indicating more than one relaxation processes.

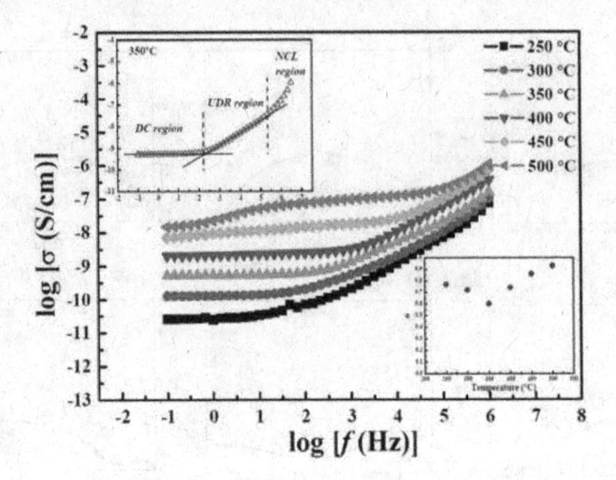

Figure 3. AC conductivity (real part) versus frequency of the TGO sample in a double logarithmic plot. The top-left inset shows the AC conductivity measured at 350°C, and the bottom-right inset shows the slope of the UDR region at various temperatures.

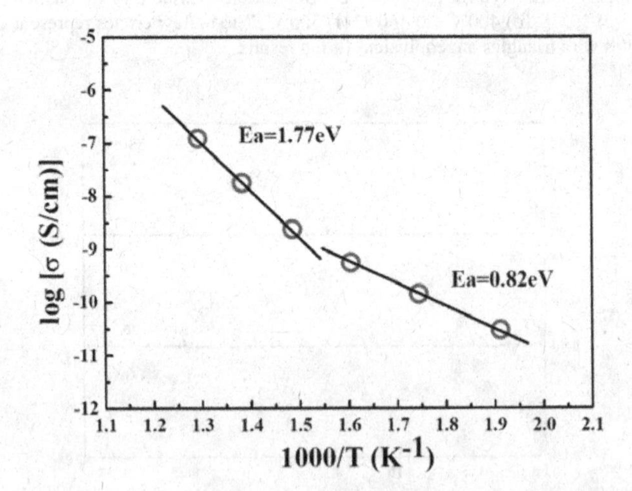

Figure 4. Arrhenius plot and activation energy values of the TGO.

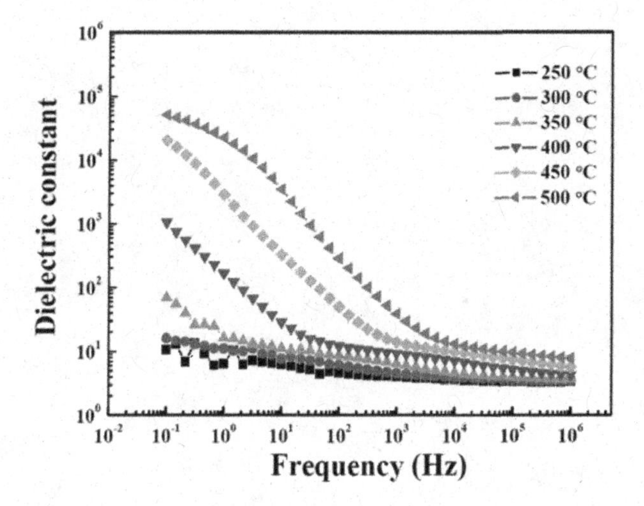

Figure 5. Frequency dependence of dielectric constant of the TGO sample at various temperatures.

MEASUREMENT OF THERMAL BARRIER COATING CONDUCTIVITY BY THERMAL IMAGING METHOD

J. G. Sun
Argonne National Laboratory
Argonne, IL 60439

Thermal barrier coatings (TBCs) are being extensively used for improving the performance and extending the life of combustor and gas turbine components. The thermal conductivity of TBCs is usually measured by destructive methods, involving separating the ceramic coating layer from the substrate and performing laser flash measurement. Nondestructive evaluation (NDE) methods would allow for direct determination of TBC conductivity on coated components and, therefore, they can be used to inspect the quality of as-processed components and monitor TBC degradation during service. This paper presents a multilayer thermal-modeling NDE method, which analyzes data obtained from pulsed thermal imaging to determine thermal conductivity distribution over the entire surface of a TBC specimen. The significance of various TBC parameters that dictate the thermal imaging data and the measurement accuracy is investigated. Experimental results obtained from an EBPVD TBC sample are presented and discussed to demonstrate the application of this method for TBC thermal conductivity measurement.

INTRODUCTION

Thermal barrier coatings (TBCs) have been extensively used on hot gas-path components in gas turbines. In this application, a thermally insulating ceramic topcoat (the TBC) is bonded to a thin oxidation-resistant metal coating (the bond coat) on a metal substrate. TBC coated components can therefore be operated at higher temperatures, with improved performance and extended lifetime [1,2]. Because TBCs play critical role in protecting the substrate components, their failure (spallation) may lead to unplanned outage or safety threatening conditions. Therefore, it is important to inspect and monitor the TBC condition to assure its quality and reliability.

Of various physical properties of a TBC material, thermal conductivity is the most important thermal property that determines the TBC quality. TBC conductivity can be measured by several methods. The most reliable and commonly used method is laser flash method on stand-alone TBC coated specimens [3]. This method however is a destructive method, requiring the TBC coat to be separated from the substrate for the measurement. Alternatively, laser flash test may also be conducted on specially-prepared TBC specimens, including the top coat, bond coat, and substrate. TBC conductivity is then determined based on multilayer material analysis for the TBC system [4]. This method still requires two-sided access of the specimen, so it cannot be used to analyze TBC coated components with variable substrate thickness. In this study, one-sided pulsed thermal imaging is investigated to determine thermal conductivity of TBC coating. It is noted that although thermal imaging has been widely applied for nondestructive evaluation (NDE) of TBCs, it has not been used for accurate prediction of physical properties for multilayer materials such as TBCs.

For a typical TBC system, a large disparity in thermal conductivity exists between the TBC and the substrate and, when TBC is delaminated with air filling the gap, between the TBC and the air. Therefore, thermal imaging can be effective for TBC characterization because it involves nondestructive measurement of thermal properties. Most of the current thermal-imaging applications have been directed for detection of TBC delamination [5]. Recently, it has also been extended for estimation of TBC thickness and thermal conductivity [6,7]. However, these methods are generally empirical or semi-empirical due to a lack of fundamental understanding of the thermal response to various TBC parameters. This paper describes a new multilayer thermal-modeling method for TBC property characterization and imaging [8]. Both theoretical and numerical methods are used to investigate the significance of various TBC parameters that dictate the thermal imaging. Results from experimental and theoretical analyses are presented and discussed.

PULSED THERMAL IMAGING FOR MULTILAYER MATERIALS

Pulsed thermal imaging is based on monitoring the temperature decay on a specimen surface after it is applied with a pulsed thermal energy that is gradually transferred inside the specimen. A schematic one-sided pulsed-thermal-imaging setup for testing a 3-layer material system is illustrated in Fig. 1. The premise is that the heat transfer from the surface (or surface temperature/time response) is affected by internal material structures and properties and the presence of flaws such as cracks [9,10]. By analyzing the surface temperature/time response, the material property and depth of various subsurface layers under the surface can be determined.

The important TBC parameters to be determined by thermal imaging include the thickness, thermal conductivity and heat capacity (the product of density and specific heat) of the top ceramic TBC layer. Optical properties of the TBC layer, which is optically translucent, will also affect the data obtained from pulsed thermal imaging. To avoid the optical translucency issue and the requirement to determine optical properties which are normally not important for TBC performance, TBCs are usually coated by a thin (graphite-based) black paint when conducting thermal imaging tests. The three TBC parameters, however, may not be independent so may not be determined individually from thermal imaging test. In comparison, the well-known laser flash method, which is based on two-sided thermal imaging, can only determine one single parameter (thermal diffusivity) from the same set of three parameters in a single-layer material. To address this issue that is essential for the developed of the multilayer thermal-modeling method, a theoretical analysis was conducted.

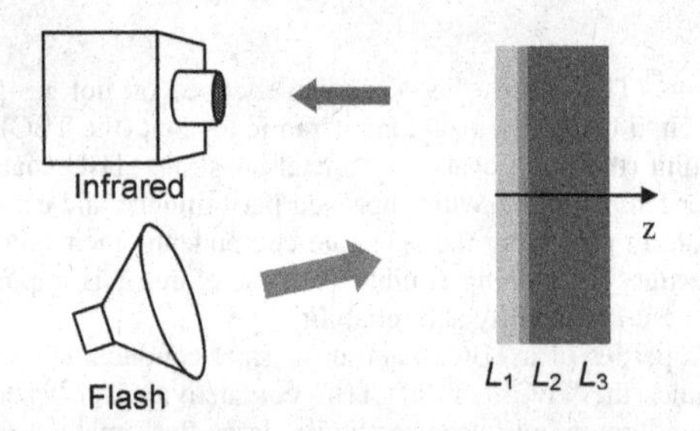

Fig. 1. Schematics of pulsed thermal imaging of a 3-layer material system.

THEORETICAL ANALYSIS OF A TWO-LAYER TBC SYSTEM

To analyze the surface temperature variation during pulsed thermal imaging, theoretical formulations of the heat transfer process need to be examined. The 1D governing equation for heat conduction in a solid material is:

$$\rho c \frac{\partial T}{\partial t} = \frac{\partial}{\partial z}\left(k \frac{\partial T}{\partial z}\right), \tag{1}$$

where $T(z, t)$ is temperature, ρ is density, c is specific heat, k is thermal conductivity, t is time, z is coordinate in the depth direction, and $z = 0$ is the surface that receives pulsed heating at $t = 0$. It is noted that Eq. (1) contains two independent thermal parameters, the heat capacity ρc and the thermal conductivity k, both are normally assumed constant in each material layer. These two thermal parameters can be converted to another two parameters, the thermal effusivity e (= $(k\rho c)^{1/2}$) and the thermal diffusivity α (= $k/\rho c$). It can be shown that any two of these four thermal parameters are independent and can be used to derive the other two parameters.

Equation (1) indicates that there are three independent TBC parameters: the thickness L, thermal conductivity k, and heat capacity ρc, when TBC translucency is not considered (TBC is normally coated by a black paint to make its surface opaque for thermal imaging test). Among these parameters, it was identified that only two parameters can be determined by pulsed thermal imaging [11]. This becomes evident when examining the solution for Eq. (1). Under ideal flash thermal imaging conditions, (i.e., instantaneous flash heating, no volume heat absorption and no heat loss), theoretical solution of the surface temperature for a two-layer material system without interface resistance can be expressed as [12]:

$$T(t) = T_\infty \left[1 + 2\frac{x_1\omega_1 + x_2\omega_2}{x_1 + x_2} \sum_{K=1}^{\infty} \frac{x_1\cos(\omega_1\gamma_K) + x_2\cos(\omega_2\gamma_K)}{x_1\omega_1\cos(\omega_1\gamma_K) + x_2\omega_2\cos(\omega_2\gamma_K)} \exp\left(-\frac{\gamma_K^2 t}{\eta_2^2}\right)\right], \tag{2}$$

where substrates 1 and 2 are for layers 1 and 2, respectively, γ_K is the K-th positive root of the following equation,

$$x_1\sin(\omega_1\gamma) + x_2\sin(\omega_2\gamma) = 0, \tag{3}$$

x and ω are defined as,

$$x_i = e_{12} - (-1)^i, \qquad e_i = \sqrt{k_i\rho_i c_i}, \qquad i = 1, 2 \tag{4}$$

$$\omega_i = \eta_{12} - (-1)^i, \qquad \eta_i = L_i/\sqrt{\alpha_i}, \qquad i = 1, 2 \tag{5}$$

with

$$e_{12} = e_1/e_2, \qquad \eta_{12} = \eta_1/\eta_2. \tag{6}$$

From Eqs. (2)-(6), it is evident that the surface temperature response under flash thermal imaging is controlled by only two parameters (in each layer): the thermal effusivity e and the parameter $\eta = L/\alpha^{1/2}$. This conclusion is significant for thermal imaging analysis of TBCs; it shows that, among the three important TBC parameters L, k and ρc, only two can be determined independently by pulsed thermal imaging.

The conclusion derived from the above theoretical analysis can be illustrated from calculations based on postulated TBC parameters. A two-layer TBC material system consisting of a ceramic top coating and a substrate is used in the calculations, with the bond coat being considered as part of the substrate because it is usually thin and its thermal properties are comparable to those of the substrate. The baseline material properties for an opaque TBC and a substrate are listed in Table 1. The calculation results are expressed in terms of the surface temperature decay slope, $d(\ln T)/d(\ln t)$, because it is more sensitive to display the details of the thermal response and is nondimensional.

Figure 2 shows the front surface temperature slope $d(\ln T)/d(\ln t)$ for TBCs with a varying parameter $\eta = L/\alpha^{1/2}$ (Fig. 2a) or effusivity e (Fig. 2b) while keeping the other parameter constant. In Fig. 2a, the slope $d(\ln T)/d(\ln t)$ is at -0.5 in the early time period because the TBCs are opaque. The slope then changes to a larger (absolute) magnitude due to an increase of thermal effusivity from TBC to substrate. It is seen that the transition time is directly related to the parameter $\eta = L/\alpha^{1/2}$, while the peak slope magnitude, at -1.07, does not change with the TBC thickness or diffusivity. Therefore, it is concluded that the parameter $\eta = L/\alpha^{1/2}$ is an independent parameter that determines the transition time for temperature-slope change, as predicted in Eqs. (2)-(6). On the other hand, the TBC effusivity e affects only the maximum slope value (see Fig. 2b), which is independent of TBC thickness L or diffusivity α. This result also indicates that the TBC effusivity could be measured directly from the maximum temperature slope value in thermal imaging data. Further, because any two of the four thermal properties listed in Table 1 are independent, the results shown in Fig. 2 further confirm that among the three TBC parameters, thickness L and two thermal properties, only two of them can be determined uniquely from thermal imaging data, as predicted from the theoretical solutions in Eqs. (2)-(6).

Table 1. Typical material properties used in calculations.

Materials	L (mm)	k (W/m-K)	ρc (J/m^3-K)	α (m^2/s)	e (J/m^2-K-s$^{1/2}$)
TBC	0.2	1.5	2.5*10^6	0.6*10^{-6}	1936
Substrate	3.0	11	3.5*10^6	3.1*10^{-6}	6205

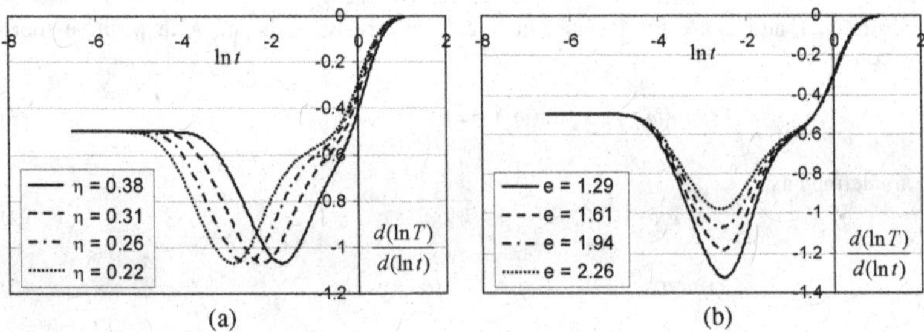

Fig. 2. Effect of TBC (a) parameter η (s$^{1/2}$) and (b) effusivity e (kJ/m^2-K-s$^{1/2}$) on surface temperature decay slope.

MULTILAYER THERMAL MODELING METHOD

The theoretical solution Eq. (2) cannot be directly used to analyze thermal imaging data because a large number of eigenfunctions needs to be calculated and a TBC system may have to be modeled by more than 2 layers. To facilitate thermal imaging analysis of multilayer materials, a multilayer thermal modeling model was developed to solve the heat transfer equation (1) numerically under pulsed thermal imaging condition [8]. In this method, a TBC is modeled by a multilayer material system and the 1D heat-transfer equation describing the pulsed thermal-imaging process is solved by numerical simulation. The numerical formulation may also incorporate finite heat absorption depth effect due to the TBC translucency and the finite flash duration effect [13]. The numerical solutions (of surface temperature decay) are then fitted with the experimental data at each pixel by least-square minimization to determine unknown parameters in the multilayer material system. Multiple parameters in one or several layers can be determined simultaneously. Among the three parameters for opaque TBCs, the TBC thickness, thermal conductivity and heat capacity, the multilayer modeling method can be configured to calculate one or two of these three parameters by setting the remaining ones at constant. This data fitting process is automated for all pixels within the thermal images and the final results are presented as images of the predicted TBC parameters.

The multilayer modeling method was used in this study to determine TBC thermal properties, conductivity and heat capacity, with known TBC thickness. The experimental sample is an as-processed EBPVD TBC specimen with a surface area of 25.4 mm in diameter (sample curtsey of Dr. A. Feuerstein, Praxair Surface Technologies, Inc,). The top ceramic coating is 0.127 mm thick, and the nickel-based supperalloy substrate has a thickness of 3.1 mm. The TBC surface was coated by a black paint for the thermal imaging test. The predicted TBC conductivity and heat capacity distributions are shown in Fig. 3. It is seen that the predicted thermal conductivity and heat capacity images are uniform. Some minor variations of the predicted properties are visible in the images, which are mostly due to the nonuniformity of the black paint and small flaws on surface. The predicted average TBC conductivity is 1.65 W/m-K, which is consistent with the typical value of such EBPVD TBCs (1.71 W/m-K) [2]. The predicted average TBC heat capacity is 3.3 J/cm^3-K, which is much higher than typical EBPVD TBCs. The reason for a higher predicted heat capacity is probably due partially to the columnar structure of the EBPVD coating so the material in the heat conduction direction is denser than the average value. In addition, current calculation did not consider the increased thickness of the top ceramic coating due to the black paint. If a thicker top coat was assumed, the predicted thermal conductivity would become slightly higher while the heat capacity be much lower. Nevertheless, these results indicate that the multilayer modeling method can be used to predict TBC parameters.

To verify the prediction accuracy, Fig. 4 compares the experimentally measured data and the corresponding predicted theoretical data for a typical pixel within the image. Although the data fit was carried out for the temperature (Fig. 4a), the fitting was also very accurate for the temperature derivative, Fig. 4b. Compared with the theoretical analysis results shown in Fig. 2, both the location and the magnitude of the temperature-derivative peak are accurately determined in Fig. 4b. The slight discrepancies of the temperature derivative data between the measured and predicted in early time period ($t < 0.01$s) was likely due to the surface roughness and effects related to the black paint, and in the later period ($t > 0.1$s) to the lower surface temperature (< 1°C) that corresponds to a lower signal to noise ratio.

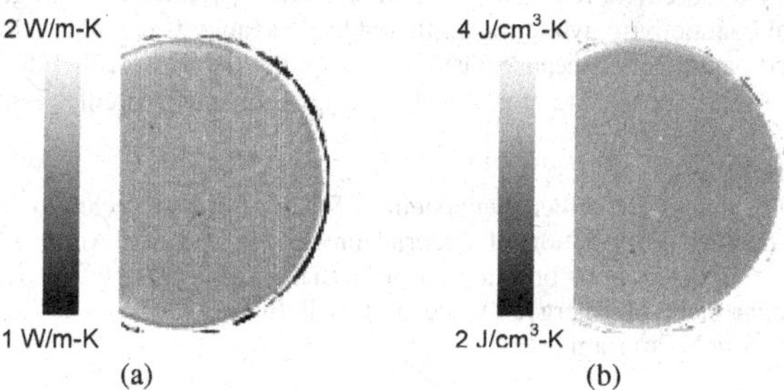

Fig. 3. Predicted TBC (a) conductivity and (b) heat capacity images of a TBC specimen.

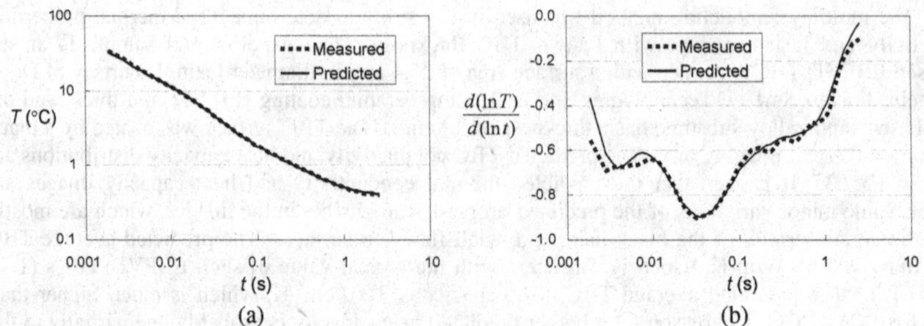

Fig. 4. Comparison of measured and predicted surface (a) temperature and (b) temperature-derivative data for a typical pixel in the Fig. 3 images.

CONCLUSION

A multilayer thermal modeling method was developed to analyze the surface temperature response for TBCs under one-sided pulsed thermal imaging condition. The significance of three TBC parameters to the surface temperature response was investigated. These parameters are TBC thickness L, conductivity k, and heat capacity ρc. Among the three TBC parameters, it was identified that only two are independent and can be determined from thermal imaging data. Specifically, these two independent parameters are e and L^2/α; the thermal effusivity e determines the peak temperature slope and the parameter L^2/α determines the transition time when temperature slope changes from -0.5 in early time to a higher magnitude. The multilayer thermal modeling method can be used to determine the thermal property distributions of the TBC layer over the entire component surface, with known TBC thickness. Measurement results for an EBPVD TBC specimen showed good model prediction of the TBC thermal conductivity, while the predicted heat capacity was higher which may be attributed to several identified parameters. Because TBC thermal properties vary with TBC service time, this NDE method may potentially be used to monitor TBC degradation and to predict TBC life.

ACKNOWLEDGMENT

The author thanks Dr. Albert Feuerstein of Praxair Surface Technologies for providing the test specimen with detailed information of material properties, and Mr. Andre Luz of Imperial College London for valuable discussions on material properties used in this study. This work was sponsored by the U.S. Department of Energy, Office of Fossil Energy, Advanced Research and Technology Development/Materials Program.

REFERENCES

1. US National Research Council, National Materials Advisory Board, "Coatings for High Temperature Structural Materials," National Academy Press, Washington, DC, 1996.
2. A. Feuerstein and A. Bolcavage, “Thermal Conductivity of Plasma and EBPVD Thermal Barrier Coatings,” Proc. 3rd Int. Surface Engineering Conf., pp. 291-298, 2004.
3. H. Wang and R.B. Dinwiddie, “Reliability of laser flash thermal diffusivity measurements of the thermal barrier coatings,” J. Thermal Spray Techno., Vol. 9, pp. 210-214, 2000.
4. B.K. Jang, M. Yoshiya, N. Yamaguchi, and H. Matsubara, “Evaluation of Thermal Conductivity of Zirconia Coating Layers Deposited by EB-PVD,” J. Mater. Sci., Vol. 39, pp. 1823-1825, 2004.

5. X. Chen, G. Newaz, and X. Han, "Damage Assessment in Thermal Barrier Coatings Using Thermal Wave Image Technique", Proc. 2001 ASME Int. Mech. Eng. Congress Expo., Nov. 11-16, 2001, New York, NY, paper no. IMECE2001/AD-25323, 2001.
6. H. I. Ringermacher "Coating Thickness and Thermal Conductivity Evaluation Using Flash IR Imaging," presented in Review of Progress in QNDE, Golden, CO, July 25-30, 2004.
7. S. M. Shepard, Y. L. Hou, J. R. Lhota, D. Wang, and T. Ahmed, "Thermographic Measurement of Thermal Barrier Coating Thickness," in Proc. SPIE, Vol. 5782, Thermosense XXVII, 2005, pp. 407-410, 2005.
8. J.G. Sun, "Thermal Imaging Characterization of Thermal Barrier Coatings," in Ceramic Eng. Sci. Proc., eds. J. Salem and D. Zhu, Vol.28, no. 3, pp. 53-60, 2007.
9. J. G. Sun, "Analysis of Pulsed Thermography Methods for Defect Depth Prediction," J. Heat Transfer, Vol. 128, pp. 329-338, 2006.
10. J. G. Sun, "Evaluation of Ceramic Matrix Composites by Thermal Diffusivity Imaging," Int. J. Appl. Ceram. Technol., Vol. 4, pp. 75-87, 2007.
11. J. G. Sun, "Thermal Imaging Analysis of Thermal Barrier Coatings," presented at the 35th Annual Review of Progress in Quantitative NDE, Chicago, IL, July 20-25, 2008.
12. D. L. Balageas, J. C. Krapez, and P. Cielo, "Pulsed Photothermal Modeling of Layered Metarials," J. Appl. Phys., Vol. 59, pp. 348-357, 1986.
13. J. G. Sun and J. Benz, "Flash Duration Effect in One-Sided Thermal Imaging," in Review of Progress in Quantitative Nondestructive Evaluation, eds. D.O. Thompson and D.E. Chimenti, Vol. 24, pp. 650-654, 2004.

THERMAL RESIDUAL STRESS IN ENVIRONMENTAL BARRIER COATED SILICON NITRIDE-MODELED

Abdul-Aziz Ali
Cleveland State University
2121 Euclid Avenue
Cleveland, OH 44115

Ramakrishna T. Bhatt
Vehicle Technology Center
US Army Research Laboratory
Glenn Research Center at Lewis Field
21000 Brookpark Rd.
Cleveland, Ohio 44135

ABSTRACT

To determine the maximum residual stresses developed during deposition of the coatings on ceramic substrate, a finite element model (FEM) was developed. Using this model, the thermal residual stresses were predicted in silicon nitride substrates coated with three environmental coating systems namely barium strontium aluminum silicate (BSAS), rare earth mono silicate (REMS) and earth mono di-silicate (REDS). A parametric study was also conducted to determine the influence of coating layer thickness and material properties on thermal residual stress. Results indicate that z-direction stresses in all three systems are small and negligible, but maximum in-plane stresses can be significant depending on the composition of the constituent layer and the distance from the substrate. The BSAS and REDS systems show much lower thermal residual stresses than REMS system. Parametric analysis indicates that in each system, the thermal residual stresses can be decreased with decreasing the modulus and thickness of the coating.

INTRODUCTION

Monolithic silicon nitride is a candidate material for next generation small engine turbine components because of its low density, high temperature strength, and high creep resistance. In addition, the processing methodology for the fabrication of complex shaped silicon nitride components is also well developed. Despite these advantages, silicon nitride is not used for flight worthy hardware of turbines because of its poor impact resistance and structural instability (surface recession) in engine operating environments [1, 2]. While impact resistance can be mitigated by component design, avoiding surface recession requires development of compatible multilayered environmental barrier coating (EBC). Problem of surface recession is not unique for silicon nitride, it also occurs in all other silicon based ceramics and SiC/SiC composites when exposed to a combustion environment containing moisture at temperatures> 1100^0C [3, 4]. To protect the SiC/SiC composites from surface recession, environmental barrier coatings (EBC) have been developed. One key example is a multilayered coating having a barium strontium aluminum silicate (BSAS) and rare earth silicate top coat [5, 6]. The BSAS and rare earth silicate based EBCs have upper temperature capabilities of ~1316^0C [6] and ~1482^0C [7], respectively for applications over 30,000 hours. In general, an EBC consists of two or more layers that each layer has its own functionality and characteristics. The layer on top of the substrate is referred to as bond coat is typically silicon, followed by an intermediate coat which is a mixture of Mullite and top coat, and then a top coat. Each one of these coating layer plays a different role in the protection scheme such as minimizing the migration of sintering of additives from substrate to the coating, avoiding sintering between layers, and reduces permeability of moisture to the substrate. These coating layers can be deposited by a variety of methods such as plasma spray (PS), electron beam physical vapor deposition (EBPVD), or slurry coating.

Literature data indicate that plasma sprayed multilayered BSAS based EBC had no effect on strength properties of SiC/SiC composites, but caused ~50% loss in strength in silicon nitride substrates [8, 9]. Various factors such as the coating process damaging the substrate surface, large tensile thermal residual stresses in the coating were suggested as mechanisms for strength degradation.

In general, when elastic and thermal properties of the coating and the substrate are different, thermal residual stresses are generated in the coated substrates depending on the thermal environment used during coating process. Depending on their nature and magnitude, thermal residual stresses can have significant effect on the strength of the substrate. To predict as well as to develop methods of lowering the thermal residual stress in coated substrates the current study was started.

The objectives of this study were several: First, to develop an analytical model to predict the magnitude of thermal residual stresses in coated substrates based upon the boundary conditions used during coating process, specimen geometry, and the thermo-mechanical properties of the coating layers and substrate; Second, apply this model to predict stresses in silicon nitride substrates coated with BSAS, Rare Earth Mono Silicate (REMS),and Rare Earth Di-Silicate (REDS) EBC systems by plasma spray; Third, to determine the influence of coating thickness and modulus on the magnitude of residual stresses.

ANALYTICAL APPROACH AND FINITE ELEMENT MODELING

The finite element method is used to determine the buildup of residual stresses for the proposed layered EBC. These analytical calculations included modeling of a beam specimen with a layer of EBC on silicon nitride substrate. The uncoated specimen dimensions were 4 by 3 by 45 mm, see Figure 1 Three different EBC thicknesses and a substrate structure are modeled. Thermal boundary conditions effects imposed by the coating application methodology are introduced into the thermal model. MSC/Patran [10] and MARC [11] finite element modeling software are used for both the geometric modeling of the specimen and as a solver for the stress state. Material properties of both, the coating and the substrate, are input through the model under linear isotropic conditions. Three dimensional model of the specimen is generated and a fine mesh through both the coating layer and the substrate interface is established. Eight-node hex type element is employed with a total average model size for each coating thickness considered of 44,000 elements and 50,000 nodes. Additional details regarding the geometry of the beam model as well as the representative thickness of each coating employed in the analyses are reported in Figure 1.

The calculations were made under linear elastic conditions where the behavior of the material is defined by the two material constants; Young's modulus and Poisson's ratio. The coated test specimen consists of three sub layers of EBC and the silicon nitride substrate. Three EBC systems namely, System-1, System-2, and System-3 correspond to BSAS, REMS, and REDS, respectively are considered. Each EBC system consists of 3 sub layers, the first is silicon layer followed by a mixture of Mullite-top layer and then a top layer. For example BSAS EBC system consists of a layer of silicon, a mixture of 50 % Mullite + 50 % BSAS layer, and then a layer of BSAS. In each EBC system two different coating thicknesses referred to in the paper as analytical Case -1 and 2 were analyzed. The analytical Case-1 which represents total coating thickness of 225 μm consists of 75 μm silicon bond coat layer, 75 μm intermediate coat layer, and 75 μm top coat layer. The analytical Case-2 which represents 325 μm thick EBC consists of 75 μm silicon bond coat layer, 125 μm intermediate coat layer, and 125 μm top coat layer.

Material property data of the silicon nitride substrate and the individual coating layer of environmental barrier coating systems required for the modeling were obtained from reference 12.

Table I shows the properties for both the coating and the substrate. Material properties for the combined coatings were derived using the rule-of-Mixtures to calculate the effective values. Thermal and mechanical boundary conditions applied during the coating application procedure as well as the material properties effects were all introduced into the analytical model. The beam geometry was constrained to match the experimental conditions and to suppress rigid body motions effects. Additionally, in order to eliminate factors such as over constraining, edge effects and other related issues; the entire specimen unit was modeled and meshed accordingly. This was a reasonably valid assumption since the analyses were performed under linear elastic conditions. Stress relaxation that may be occurring in the coating during plasma spray deposition was neglected in the model. All the nodes at the specimen base or bottom surface were constrained not to move along the Z-axis and an additional node at the bottom corner section was constrained not to move along X, Y and Z axes to suppress rigid body motion was also retained. This simulated the fact that the specimen is situated on flat surface while being exposed to thermal coating applications.

Thermal environment was imitated by applying an initial temperature of 1200 °C to the entire structure followed by a cool down to room temperature of 21 °C. This represented the thermal setting experienced by the beam specimen while being plasma sprayed. Subsequently, residual stresses in the coating, the substrate and the combined structure are all determined and evaluated.

Table I. Physical, thermal and mechanical properties of silicon nitride substrate and standalone plasma sprayed
coatings [12]

Material	**ρ gm/cc**	**Bend Modulus, GPa**	**Bend Strength, MPa**	**Thermal Expansion, α E-6/°C**	**Poisson' Ratio, γ**
Substrate, Silicon nitride	3.4	318	595	3.8	0.18
Plasma sprayed coatings					
Mullite	2.9	45	28	5.8	0.17
Rare Earth-Mono Silicate (REMS)	7.9	91	57	8.1	0.18
Silicon	2.3	97	40	4.5	0.21
Rare Earth Disilicate (REDS)	7.9	90	60	4.8	0.18
Barium Strontium Aluminum Silicate (BSAS)	3.2	32	28	5.6	0.19

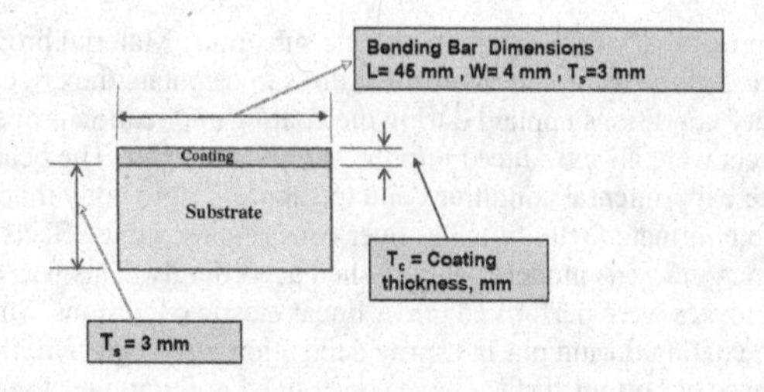

Figure 1. Test Bar Specimen Geometry (not drawn to scale)

RESULTS AND DISCUSSION

Results obtained from the finite element analyses are illustrated in Figures 2 to 6. Figure 2 shows the stresses in the bar specimen due to the thermal loading applied for both the coating and the substrate. These results are for system-1 with coating sequence as Silicon-(Mullite+BSAS)-BSAS and case-1 respectively. The coating thickness of each layer is 75 μm. The coating and substrate thickness is designated by the symbols t_c and t_s and they represent the incremental thickness of the coating and the substrate respectively, while T_c and T_s are the total thickness of each entity. The location at the substrate/coating interface is represented by a ratio of zero for t_c/T_c, while a ratio of unity represents maximum coating thickness. Similar convention is used concerning the ratio of the substrate thickness arrangement .i.e. t_s/T_s corresponds to maximum substrate thickness In figure 2(a) the stresses along the thickness, Z-axis, and along the X-Y axes are shown as a function of the normalized distance. The normalized distance is defined as the ratio of the length increment divided by the total thickness. It is reported from the data presented that the X-Y stresses are much higher than the through thickness stress; in fact the through thickness stresses are nearly negligible as projected. Figure 2(b) represents the stresses along the X, Y and Z axis through the substrate. Their magnitude is relatively small.

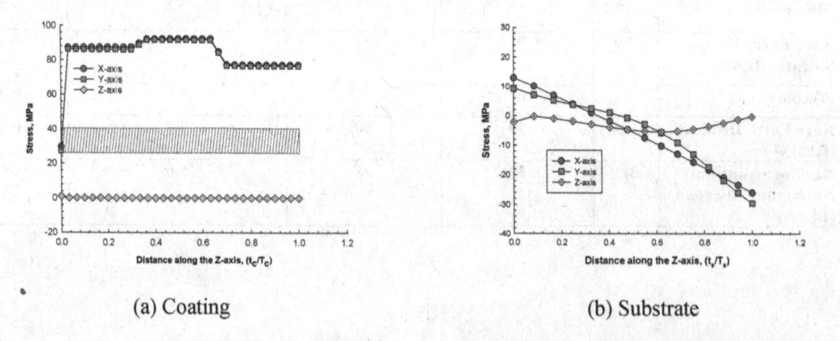

(a) Coating (b) Substrate

Figure 2. Predicted variation of thermal residual in-plane stress versus normalized distance for System 1, Case 1. The hatched area in the figure indicates typical flexural strength range of standalone coating layers.

Similarly, the predicted stresses for the System-1, analytical Case-2 are shown in Figures 3(a) and 3(b). The data shown in Figure 3(a) correspond to the in-plane and the through thickness stresses in the coating. In this case, the bond coat thickness is much smaller than the subsequent two coats.

Comparison of figure 2(a) and 3(a) indicates that following: first, the X and Y stresses in the intermediate layer of the coating in both cases are relatively greater than either the silicon bond coat layer or top BSAS layer.; second, the X and Y stresses in the silicon bond coat for case 1 are nearly the same as that in the intermediate coat, but for case 2, X and Y stresses in the silicon bond coat are almost zero.; third, in general X and Y stresses for Case 2 are ~ 25% lower than case 1 which suggests that by manipulating relative thickness of individual layers as well as increasing overall thickness of the coating, it is possible to decrease thermal stresses.

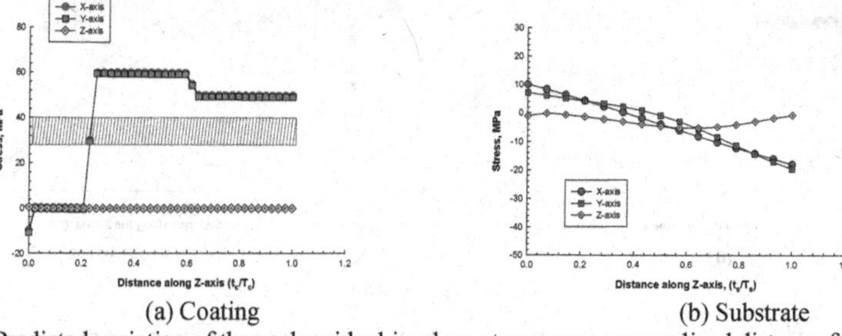

(a) Coating (b) Substrate

Figure 3. Predicted variation of thermal residual in-plane stress versus normalized distance for System 1, Case 2. The hatched area in the figure indicates typical flexural strength range of standalone coating layers.

Figure 4(a) and 4(b) show the results obtained for System-2, Case-1. It is very clear from Figure 4(a) that the in-plane stresses are substantially larger than those experienced by system-1 under the same conditions of Case-1, Figure 2(a). They are an order of magnitude higher compared to Case1/System-1 data. Similarly, the stresses through the substrate under system-2, Case-1 also showed slightly higher stresses, nearly doubles in magnitude, compared to those obtained under system1, case1. However, their extents constitute no significance on the structural behavior of the silicon nitride, they remain relatively small.

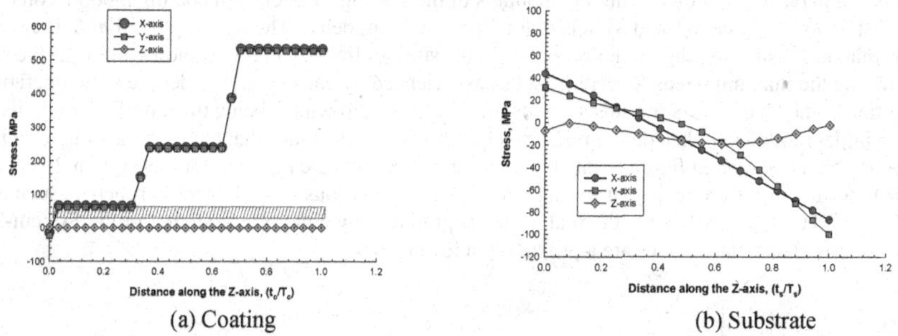

(a) Coating (b) Substrate

Figure 4. Predicted variation of thermal residual in-plane stress versus normalized distance for System-2, Case-1. The hatched area in the figure indicates typical flexural strength range of standalone coating layers.

Figures 5(a) and 5(b) represent results obtained under the same thermal conditions for the System 3, Case-1. The elastic properties of system-3 are similar to that of System-2, but the thermal expansion of System-3 is nearly half that of System-2.

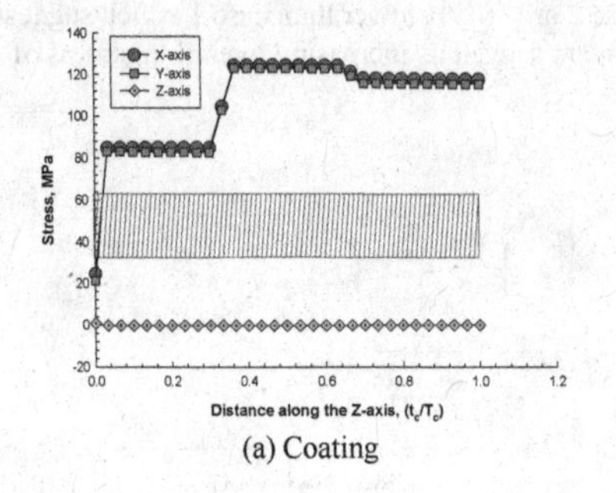

(a) Coating

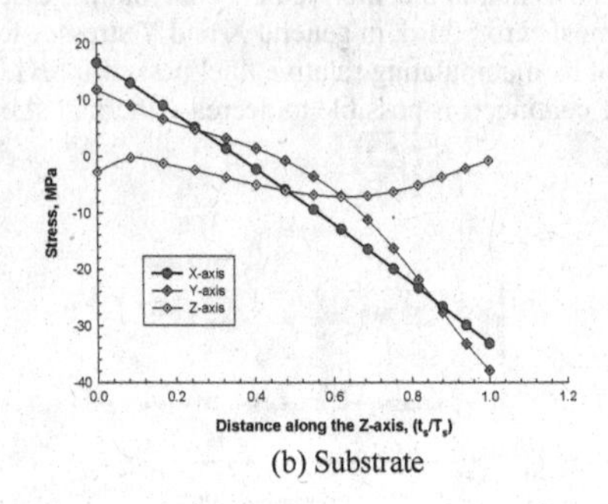

(b) Substrate

Figure 5. Predicted variation of thermal residual in-plane stress versus normalized distance for System-3, Case-1. The hatched area in the figure indicates typical flexural strength range of standalone coating layers.

Figure 5 shows that X and Y stresses in this system are much lower than those obtained for System-2, Case-1 and are very comparable to the results obtained for the System-1, Case-1. Similarly, the stresses reported through the substrate are equally low and insignificant. Comparison of modeling data for all three systems suggests that both system-1 and System-3 are better systems for development of EBC for silicon nitride.

In plasma spray deposition, it is possible to control porosity in the coating by varying processing parameters, thus varying the modulus of the coating. Influence of coating modulus on the X and Y stresses for System-1 and System-2 for Case-1 was modeled. The results plotted in figure 6 show that decreasing coating modulus decreases the stresses linearly Figure 6 shows an exclusive chart illustrating the maximum tensile residual stress experienced by each system under the same operating conditions. The chart shows the stress versus the modulus ratio with E being the modified modulus and E_0 the initial modulus of the coating respectively. This is demonstrates that the modulus influence is obvious. It is all shown in figure 6; much higher tensile stresses are reported for the System-2 versus those reported for the System-1. It further confirms that variations in the material properties such as modulus and CTE will affect the thermal response produced by either the System-1 or the System-2. Lower thermal residual stresses are a product of a lower CTE.

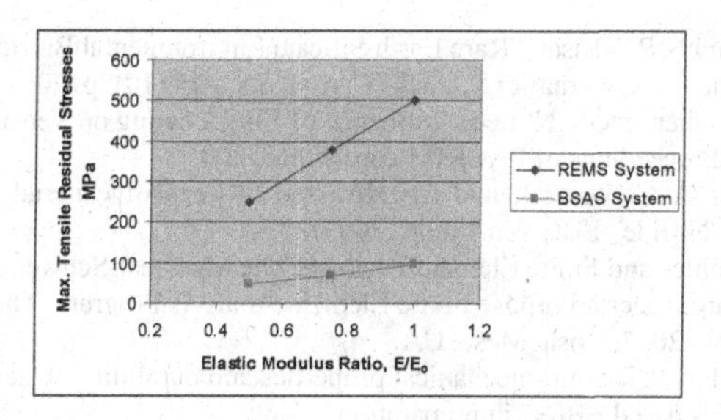

Figure 6. Influence of coating modulus on thermal residual stress for silicon nitride coated with REMS and BSAS systems, Case-1.

CONCLUSIONS

A finite element model was developed to predict in-plane and through-the-thickness stresses in a rectangular beam of coated ceramic substrate simulating the experimental environment conditions experienced during the coating process and using the constituent properties of the coating and the substrate. Thermal residual stresses developed in silicon nitride substrate coated with three environmental barrier coating systems- BSAS, REMS, REDS- (Systems 1, 2 and 3) and for two overall coating thicknesses (analytical Cases 1 and 2) were modeled. Influence of coating modulus and thermal expansion on the thermal stresses was studied. Majors finding are the following.

(1) The z-direction (through thickness) stresses in all three systems for case-1 are small and negligible, but maximum in-plane stresses can be significant depending on the composition of the constituent layer and the distance from the substrate. (2) The in-plane thermal residual stresses for System-2, Case-1 is much higher than those for System-1 or System-3, Case-1.
(3) Reducing the elastic modulus and the coefficient of thermal expansion will result in lowering stress response.

REFERENCES

[1] O. Jaimenez, J. McClain, B. Edwards, V. Parthasarathy, H. Bhageri, and G. Bolander ASME paper-1998-529
[2] T. Itoh, Y.Yoshida, S. Sasaki, M. Sasaki, and H. Ogita, "Japanese Automotive Ceramic Gas Turbine Development," p283-303, V 1, Eds M. van Roode, M.K. Ferber, and D.W. Richerson, The American Society of Mechanical Engineers, Three Parks Avenue, New York, NY 10016, 2002.
[3] E.A. Gulbransen and S.A. Jansson, "The High-Temperature Oxidation, Reduction, and Volatilization Reactions of Silicon ad Silicon Carbide," *Oxidation of Metals*, **4**[3] p181-201, (1972)
[4] P.J. Jorgensen, M.E. Wadsworth, and I.B. Cutler, "Oxidation of Silicon Carbide," *J. Am. Ceram. Soc.*, **42** (12): p613-616 (1959).
[5] K.N. Lee, *Surface Coating Technology*, p133-134, 1-7, 2000.
[6] K.N. Lee, D.S. Fox, J.I. Eldridge, D. Zhu, R.C. Robinson, N.P. Bansal, and R.A. Miller, *J. Am. Ceram. Soc.*, **86** (8): p1299-1306, 2003.

[7]K.N. Lee, D.S. Fox, and N.P. Bansal, "Rare Earth Silicate Environmental Barrier Coatings for SiC/SiC Composites and Si_3N_4 Ceramics," *J. Eur. Ceram. Soc.*, **25** (10): p1705-1715, 2005.
[8] R.T. Bhatt, G.N. Morscher, and K.N. Lee, "Influence of EBC Coating on Tensile Properties of MI SiC/SiC Composites," Proceedings of PACRIM conference, 2005
[9] H.T. Lin, M.K. Ferber, A.A. Werezak, and T.P. Kirkland, "Effects of Material Parameter on Strength of EBC Coated Silicon Nitride" EBC Workshop (2003).
[10] MSC/PATRAN Graphics and Finite Element Package. The MacNeal-Schwendler Corporation, 2007, Costa Mesa, CA. [11] Marc General Purpose Finite Element Analysis Program, The MacNeal-Schwendler Corporation, 2007, Costa Mesa, CA.
[12] R.T. Bhatt and D.S. Fox, "Thermo-mechanical properties and durability of stand alone constituent layers of BSAS and RES based EBCs" in preparation

FRACTURE MECHANICAL MODELLING OF A PLASMA SPRAYED TBC SYSTEM

Håkan Brodin[1,2], Robert Eriksson[2], Sten Johansson[2], Sören Sjöström[1,2]
[1]Linköping University, 581 83 Linköping, Sweden
[2]Siemens Industrial Turbomachinery AB, 612 83 Finspång, Sweden

ABSTRACT

A thermal barrier coating (TBC) system subjected to thermal cycling will develop a microcrack pattern near the interface between the metallic bond coat and the ceramic top coat. These small cracks link up and form internal TBC delaminations during repeated heating / cooling. After a longer time period, the internal delamination cracks will form a larger spallation damage, where the TBC is detached from the underlying material. Since cracks are initiated in multiple sites of the thermal barrier coating, the damage is initially considered to be governed by local stress conditions. The purpose of the present work is to compare experimental data with predictions of a physically based fatigue life model. The present study has been performed on plasma-sprayed TBC's where the interface geometry has been varied. In the present work, calculation of fatigue life is done for a number of cases under thermal fatigue loading. Different interface geometries are compared in order to understand the influence of variations in the TC/BC interface roughness on oxidation behaviour and thermal fatigue life. Thermal fatigue tests indicate that an increased surface roughness is beneficial from a fatigue life point of view.

INTRODUCTION

Thermal barrier coatings (TBC) are applied on hot section components in aero engines and land-based gas turbines [1,2]. The main reason for applying a thermal barrier coating is to reduce temperature and, hence, reduce the risk for oxidation and creep damage and increase the component fatigue life. In order to fully make use of the benefits from a thermal barrier coating, reliable life prediction methods are required, and several proposals have been suggested in the literature [3-5].

TBC's are normally built up of at least two to three layers on a base material (BM), where each layer has one or more unique functions. In the classic TBC system two layers are considered, the bond coat (BC) and the top coat (TC). Oxidation and corrosion protection can be provided by the bond coat together with good bonding to the ceramic top coat. Bond coats are most commonly based on the MCrAlX group of alloys, where M indicates additions of nickel, chromium, cobalt and in some cases also iron. X denotes small additions (about 1 wt%) of elements such as yttrium, hafnium or tantalum. Partially stabilised zirconia is normally chosen as top coat, since this will give a coating with as good crack growth resistance as possible together with good thermal insulation performance. Stabilisation can be done with ceramics such as CaO, MgO or CeO, however, 6-8wt% Y_2O_3 is the most common choice [6].

The life limiting damage mechanism of thermal barrier coatings is mainly the detachment of the ceramic top coat from the metallic bond coat [7]. Delamination cracks will form due to development of a thermally grown oxide (TGO) and presence of thermal stresses at the TC / BC interface. Since aluminium is present in all MCrAlX coatings, the TGO is normally alumina. However, with time the aluminium is consumed and other oxides will form. These oxides are either of spinel type, nickel oxide or chromia [8-10]. Depending on the magnitude of the thermal stresses and the growth rate of the TGO, the delamination cracks develop either in the TGO (at the TC/BC interface) or in the TC near the TC/BC interface [11,12]. After repeated thermal and/or mechanical loading the delamination cracks link up to form a larger spallation crack, which, in turn, leads to local failure of the thermal barrier coating caused by buckling of the coating [13,14].

A fracture mechanics based lifing methodology for assessment of spallation fatigue life has previously been proposed by Brodin and Jinnestrand [15,16]. In the present paper, the aim is to compare the life prediction model to a series of TBC fatigue tests where the geometry of the TC / BC interface has been varied. The TC and BC compositions were kept constant and all tests were performed on one type of base material.

MATERIALS

The present study is performed on a series of four TBC systems with vacuum plasma sprayed (VPS) bond coats and air plasma sprayed (APS) top coats. Spraying in absence of air provides a homogeneous bond coat with a minimum of pores and internal oxides. As bond coat a NiCrAlSiTaY alloy was used. The bond coat was sprayed to different surface roughness in the four cases in order to evaluate the effect of surface roughness on TBC spallation fatigue life. Therefore the spray parameters were modified so that the desired surface roughness variation could be accomplished. The main factor that was modified was powder size distribution. In all four cases the top coat was identical, zirconia partially stabilised by 7wt% yttria (7Y-PSZ). Top coat porosity was kept constant in all four test series. All spray tests were performed on rectangular coupons (30x50x5mm) manufactured from the solid solution strengthened nickel-base superalloy Haynes 230. Nominal compositions (wt%) of base material, bond coat and top coat are shown below, Table 1.

Table 1. Nominal composition of base material, bond and top coat material in the present study (wt%).

	Ni	Cr	Co	Al	Mn	Mo	Si	W	Fe	Ti	Ta	Cu	La	B	Y	Y_2O_3	ZrO_2
BM	bal.	22	5	0.3	0.7	2	0.5	14	1.5	0.1	–	0.5	0.005	0.007	–	–	–
BC	bal.	25	–	5	–	–	2.6	–	–	–	1	–	–	–	0.6	–	–
TC	–	–	–	–	–	–	–	–	–	–	–	–	–	–	–	7	bal.

Bond coat and top coat are sprayed to a nominal thickness of 200µm and 350µm, respectively (average values).

CHARACTERISATION AND TESTING

Damage development in thermal barrier coatings is, as mentioned above, dependent on a number of factors where bond coat oxidation behaviour, top coat porosity, top coat thickness and bond coat surface roughness can be mentioned. Jinnestrand and Sjöström performed a series of 3D FE simulations of the stress state at the TC/BC interface during growth of the TGO [17]. Jinnestrand and Sjöström showed that the location of maximum stress along the wavy TC/BC interface changed from peak to valley during TGO growth up to 8µm. The same trend is also later shown by Casu et.al. [18].Elsewhere in literature, it is also suggested and verified that delamination cracks are normally initiated at the peaks of the TC/BC interface [19,20]. Detection of delamination crack growth cannot easily be non-destructively performed in APS top coats, since damage development is occurring throughout the entire TC and TGO. Brodin [21,22] and Brodin et.al. [23] have, however, performed damage development investigations where damage was measured on cut and polished cross sections of interrupted tests. From such investigations a measure of interface damage could be established, where cracks either form and grow at the TC/BC interface (interface cracks), form and grow in the TC near the TC/BC interface or form at the TC/BC interface and grow partly at the interface, partly in the TC near the interface, see Equation 1.

$$D_m = \frac{\sum_i l_i^{TGO} + \sum_j l_j^{TC} + \sum_k l_k^{TC/TGO}}{L_m} \qquad (1)$$

In Equation 1 D_m is measured damage, l_i^{TGO}, l_i^{TC} and $l_i^{TC/TGO}$ are lengths of individual cracks in the TGO, in the TC and partly in TC, partly in TGO respectively. L_m is the total measurement length, see also Figure 1. When actual spallation cracks reaches a size of a few mm it is possible to detect the damage visually, since areas with detached TBC will have a different temperature during cycle transients than areas with intact TBC due to poor heat transfer from detached areas. Hence, these areas will glow at high temperature during the cool-down part of a thermal cycle. This method has been used in fatigue tests performed here, since the number of cycles to failure has been of main interest for the present work.

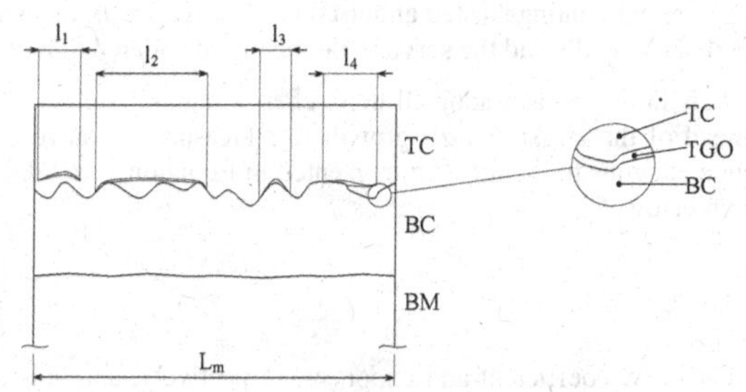

Figure 1. Damage development in a thermal barrier coating system

Measurement of surface roughness has been done on LOM micrographs of polished TBC cross sections by an in-house Matlab routine. High-resolution micrographs of the interface are combined to an adequately long interface length, digitalised and the centre line of the TC/BC interface profile identified. Eight pictures of each interface are evaluated and each micrograph typically covers 1-1.3mm interface length. After digitalising and centre line recognition, the waviness, w_a, of the interface profile is detected, where long wavelengths (larger than 650 µm) are removed. The long wavelengths can be seen as a curvature of the base material rather than waviness of the bond coat.

Fatigue testing of thermal barrier coatings has been performed as thermal cyclic fatigue (TCF) testing. TBC coated specimen have been thermally cycled between 100°C and 1100°C. Average heating and cooing rates were set to 1,7°C/s. The furnace was set to the maximum temperature while the sample holder was moving between the furnace and n external cooling nozzle. Cooling from maximum temperature has been done by cooling of the TBC surface by compressed air. When 20% of the TC/BC interface was detached, each test was considered to be failed. 20% detachment by spallation is, thus, a measure of large continuous spallations (mm-cm size), rather than small (µm-mm size) delamination defects. For comparison purposes specimens have also been isothermally oxidised at 1100°C for 1000h in order to evaluate possible differences in oxidation rate due coating manufacturing influence on BC microstructure and composition, factors that have impact on the TBC fatigue life [24].

Evaluation of the coating microstructures was performed with light optical microscopy (LOM) and a Hitachi SU-70 high resolution field emission gun scanning electron microscope (FEG-SEM).

Analysis of the BC composition and variations was performed by energy dispersive spectroscopy (EDS) in the SEM.

TBC LIFE MODELLING

The fatigue life formulation used in the present work is based on the Paris law formulation, which is widely used to relate the mode I stress intensity factor to the crack growth rate in a homogeneous material. A modification of Paris law is needed, because the delamination crack path in the TBC system does not follow the path where $K_{II} = 0$. Instead the direction of crack growth is parallel to the TGO layer, giving a crack that is influenced both by pure opening and shear loads. In the model, cracks are assumed to form at each ridge and grow down the valleys as shown for two neighbouring cracks below in Figure 2. Each crack has an extension $2a_p$ and can reach a total length of $2L_c$. When $a_p = L_c$, D equals unity and the ceramic TC is totally detached from the bond coat, since all neighbouring cracks are now linked up to each other. In order to set a suitable failure criterion, the value $D_{crit} = 0.85$ was previously proposed for a similar APS TBC system. The main reason for using a damage failure limit is because normally are not coatings tested and used until the entire TC is detached. Rather is the normal scenario that TC locally spalls and the service life then is considered to be consumed.

In the modified Paris law approach adopted here, energy release rate is used for calculation of spallation damage. Instead of the classic *da/dN* growth rate measure, we have here chosen to use a development of damage analogous to the measure presented in Equation 1, *dD/dN*. The resulting Paris law formulation is shown below, Equation 2.

$$\frac{dD_c}{dN} = C(\lambda \Delta G)^n \tag{2}$$

Here C and n are the Paris law coefficient and exponent, respectively, and λ is a coefficient to take mode mixity into account. λ is described in Equation 3 by an expression based on previous work by Suo and Hutchinson [25]

$$\lambda = 1 - (1 - \lambda_0)\left(\frac{2}{\pi}\tan^{-1}\left(\frac{\Delta K_{II}}{\Delta K_I}\right)\right)^m \tag{3}$$

where λ_0 and m are fitting parameters for the crack growth rate change due to the mode mixity (combination of opening/shear loading). Note that λ takes the value 1 when $\Delta K_{II}/\Delta K_I \to 0$ and the value λ_0 when $\Delta K_{II}/\Delta K_I \to \infty$.

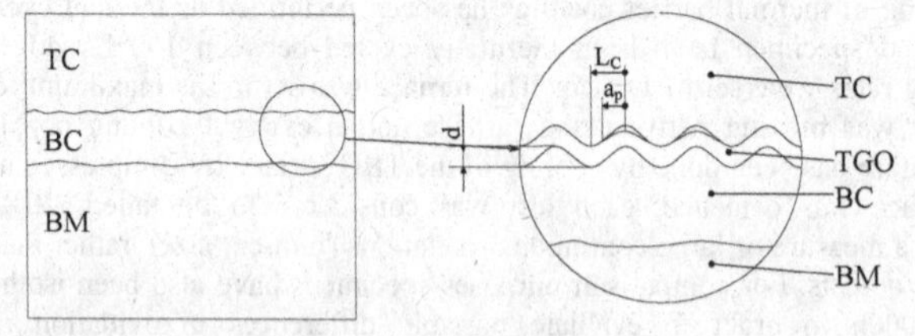

Figure 2. Definition of dimensions used for the fracture mechanical analysis.

Energy release rate and stress intensity factors are retrieved from FE analyses of a small cell covering three neighbouring cracks. Damage is assumed to develop at low temperature, since the stresses in the TBC will relax at high temperature according to the model assumptions. The assumption of damage development at low temperature has been experimentally verified by Renush et.al. with aid of acoustic emission [26]. By performing the analysis for a number of crack sizes $0 < a_p < L_c$, oxide thickness levels and external load cases, it will be possible to develop a fracture mechanical database where fracture mechanical data (G, K_I and K_{II}) can be retrieved for any case by interpolation. $D_c(N)$ can then be computed from Equation 2 by numerical integration. If the fatigue limit N_f is defined as the number of cycles where the critical damage D_{crit} is reached, then the damage development is given by Equation 4.

$$D_{crit} = \int_0^{N_f} f\left(\Delta G, \frac{\Delta K_{II}}{\Delta K_I}, \dots\right) dN \qquad (4)$$

Since a fracture-mechanical computation always assumes an existing crack, the average initial crack size a_0 must be known, since a_0 corresponds to an initial damage D_0. Such initial cracks are usually present as a result of the production process and are typically in the μm range or less for the present type of materials. Here D_0 is estimated to the order of 0.05.

The coupling between experimental and modelling data is to simply equal the measured damage to the model damage measure, $D_m = D_c$. By comparing the results of the thermal cyclic tests with the results of the finite element model for the same conditions, parameters C, n, m and λ_0 of the model can be calibrated for a given TBC system. However, it will be possible to apply the model to similar systems, given that the fracture mechanical database (containing fracture mechanical G, K_I and K_{II} data) is updated accordingly.

RESULTS

Four TBC systems, all with the same TC and BC composition but with variations in BC surface roughness, are included in the present study. Variation of interface roughness is performed by altering powder size distribution during spraying of the four different bond coats. Micrographs of the as-coated microstructure are shown below in Figure 3 for the four tested TBC variants.

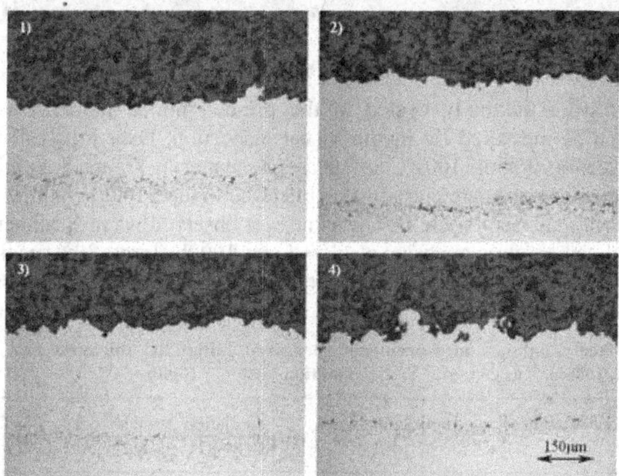

Figure 3. Microstructures of the four tested TBC variants 1-4.

In Figure 4 is shown the microstructure and corresponding EDS maps for aluminium, nickel chromium, silicon and tungsten in the as-deposited type 1 bond coat. The microstructure is homogeneous except for chromium that seems to be locally enriched in small islands within the bond coat. It is also obvious that the aluminium concentration is locally higher at the TC / BC interface compared to the BC bulk. From the SEM analysis, it was not possible to detect any homogeneous alumina scale at the TC/BC interface. However presence of β-NiAl is observed in all coating systems and most easily observed near the TC/BC interface, hence explaining the Al-peak near the TC/BC interface. At the BC/BM interface, the local aluminium-rich areas are explained by residual grit blasting media from the manufacturing process.

The surface roughness has been measured according to the previously described procedure. Data are presented below in Table 2 together with measurements of initial bond coat thickness and oxide thickness data for static oxidation in air. Static measurements are shown, since the four fatigue test cases are not interrupted at the same time, but at a corresponding global spallation damage level of 20% TBC flaking.

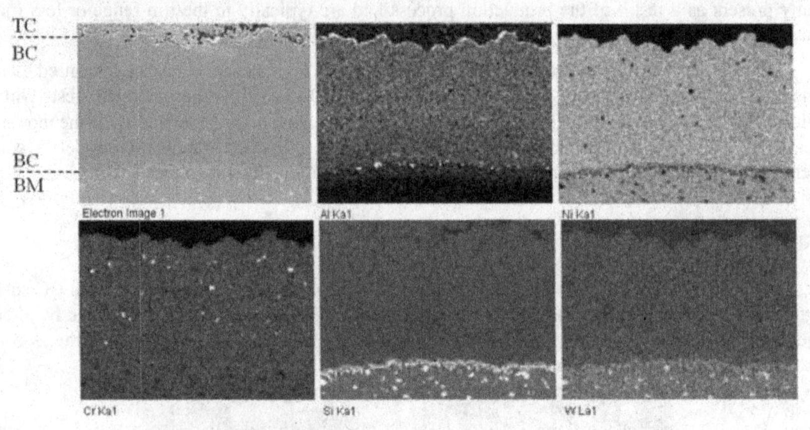

Figure 4. EDS map of the as-deposited TBC system type 1 bond coat.

The specimens were thermally cycled to the pre-determined failure criterion, 20% visible macroscopic spallation as measured by in-situ visual inspection. Data for TBC spallation life after thermal cycling between 100 and 1100 °C in air are presented in Figure 5 as a function of surface roughness w_a. Spallation damages are classified as interface failures (black failures) in all four coating systems after thermal fatigue, see Figure 6. The same was observed when specimens of the same type were subjected to high temperature exposure at 1100°C for 1000h. Two of the specimens, type 1 and 2, showed spallation damage after 1000h. At this point of time the TGO thickness was measured to approximately 10μm for all four coating systems as reported below, Table 2.

Table 2. Data for surface roughness measurements, measured initial BC thickness and TGO thickness after exposure in air, 1100°C, 1000h. 1) as coated, 2) static oxidation 1100°C / 1000h

TBC system	Surface roughness, w_a [μm] 1)	BC thickness [μm] 1)	Spallation damage after static HT exposure 2)	TGO average thickness [μm] 2)
1	7,8868	181	Yes, black failure	9,7
2	6,8345	293	Yes, black failure	9,6
3	7,7942	265	No	10,3
4	10,66	224	No	9,5

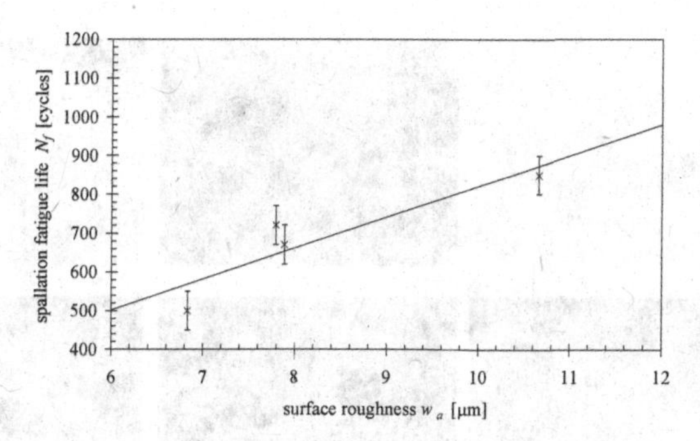

Figure 5. Spallation fatigue life as a function of surface roughness $N_f = f(w_a)$ for the four coating systems used in the present study.

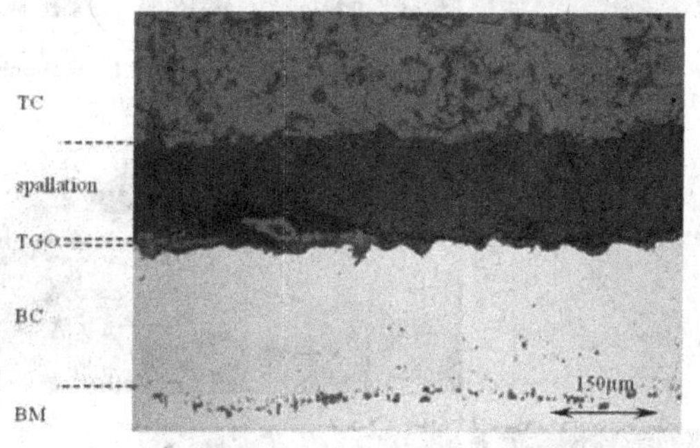

Figure 6. Failure is caused by growth of interface cracks in all four coating systems. Above is shown the interface spallation of TBC system type 1 as an example. Some Y-PSZ particles are visible on the TGO surface.

Figure 7 shows the coating type 1 TBC microstructure and EDS maps for aluminium, nickel chromium, silicon and tungsten after high temperature exposure. The aluminium is mainly concentrated to the TC/BC interface as alumina. Also at the BC/BM aluminium-rich islands are detected. These islands were previously determined as stemming from the manufacturing process. In the substrate precipitations of silicon and chromium can be observed. An interdiffusion zone has developed with a width corresponding to BC thickness, i.e. in the range of 200μm. Tungsten in the base material seems to be more finely dispersed after high temperature exposure in comparison to the initial microstructure.

In Figure 8 is the typical TGO microstructure and composition as measured by EDS for aluminium, yttrium, nickel and chromium after the thermal fatigue test shown. The thermally grown oxide consists of alumina with small clusters of aluminium and yttrium, indicating presence of aluminium-yttrium-oxides. At the TC/TGO interface is a thin band of chromia observed. Presence of oxides other than alumina at the TC/BC interface is an indication that the aluminium reservoir in the BC is lost and that the interface is degrading. Due to the formation of a voluminous oxide, in this case chromia, the bond strength at the TC/TGO interface will be deteriorated [27], and, hence, TC/TGO spallation cracks can propagate more easily.

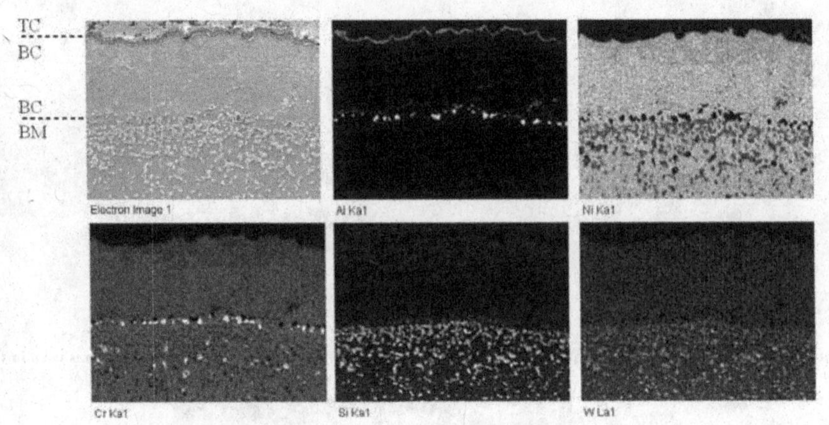

Figure 7. Microstructure after thermal fatigue conditions. The BC is depleted from aluminium and silicon mainly together with chromium has formed an interdiffusion zone in the base material.

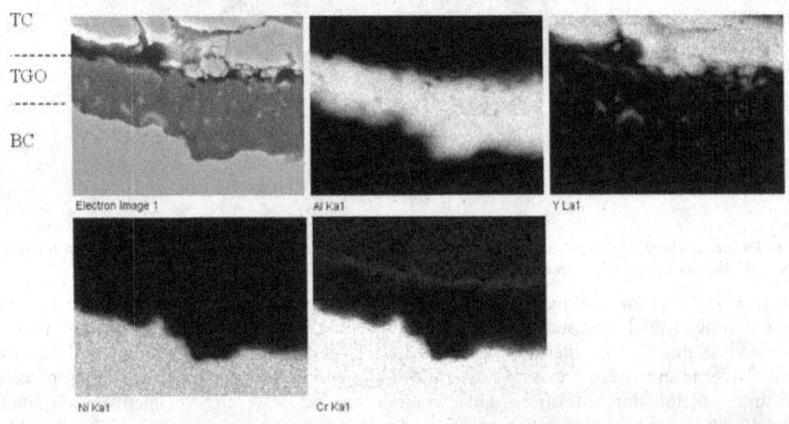

Figure 8. EDS analysis of the TGO of TBC system number 4 after thermal cycling. The TGO consists mainly of alumina. An external chromia layer is about to form. In the figure are also stringers of a mixed yttria-alumina oxide within the TGO identified.

The aim of the present work is to couple experimental findings to model predictions of lifing data. A model for this type of damage in plasma-sprayed thermal barrier coatings has been proposed by Brodin and Jinnestrand, as described previously. In the work by Brodin and Jinnestrand, the model of Paris-law type was calibrated to a set of air-plasma sprayed test coupons exposed to thermal fatigue. The resulting fit is presented below for reference, Figure 9.

By using the model after fitting, the fatigue life for a number of cases was predicted. Here the interface between top- and bond coat has been assumed to be represented by a sine wave, where the curve is assumed to have a wavelength $2L_c$ of 140μm and an amplitude $d/2$ of 5, 10 and 15μm for the cases in question.

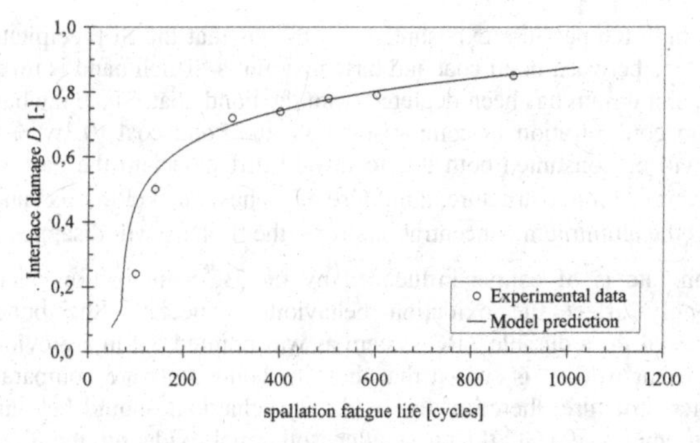

Figure 9. Fit of the model proposed by Brodin and Jinnestrand to fatigue life data for an air-plasma sprayed TBC system.

The resulting calculated damage development D_c is shown in Figure 10. The d/L measure can of course be interpreted as a surface roughness measure. If the thickness (i.e. the sinusoidal geometry) is integrated over half a wavelength (L) and the area is distributed as a constant thickness over L, then the measure is comparable to the measured surface roughness. Corresponding values for 10/70, 20/70 and 30/70 μm/μm interface geometries are w_a = 3, 6 and 12μm, i.e. the 20/70 and 30/70-profiles are closest to the four geometries in the present test series. The difference between the life prediction to reach D_c = 0.85 for the two profiles is 11% longer life for the more coarse interface.

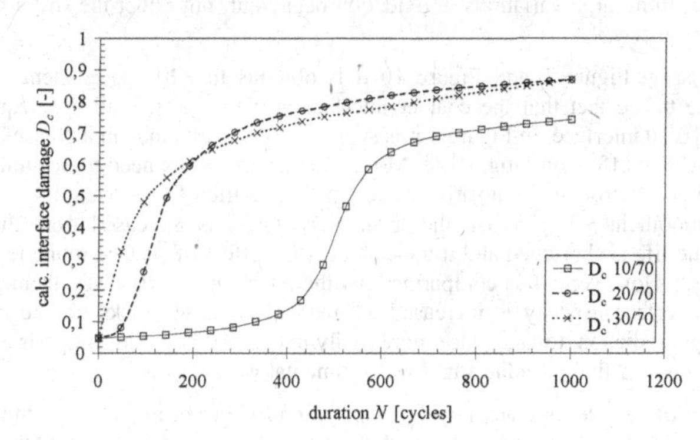

Figure 10. Calculated damage development D_c as a number of elapsed cycles N, $D_c = f(N)$. Calculations are performed for three cases, d/L = 10/70, 20/70 and 30/70μm.

DISCUSSION

The micrographs and EDS maps shown for coating system type 1 can very well represent the initial and ex-serviced microstructures of the four coating systems. Initially the bond coat has a high chromium content with isolated islands of precipitated α-Cr. With time at temperature, the chromium content is lowered, chromium precipitates in the base material. After TBC failure the EDS analysis reveals precipitation of Cr-rich particles in a diffusion zone extending approximately 200μm from the bond coat /base material interface into the base material. Silicon precipitates are present in the virgin base material. After high temperature exposure, it is obvious that the Si precipitates are increased in number. At the interface between bond coat and base material a Si-rich band is revealed. Furthermore, it is obvious that the aluminium has been depleted from the bond coat. Since the base material initially has a low aluminium concentration in comparison with the bond coat (0.3wt% to be compared to 5wt%), aluminium will be consumed both due to inward diffusion into the base material and due to oxidation. In the as-sprayed microstructure, a mixture of γ-phase (nickel matrix) and β-phase (NiAl) is observed. As soon as the aluminium concentration drops, the β-phase will disappear.

TGO oxidation rate is of course influenced by the BC composition, since the constituents available will strongly affect the oxidation behaviour. Especially the bond coat aluminium concentration is essential for a durable TBC system as was pointed out in a previous work by Haynes et.al. [24]. In the present work it is shown that the four bond coats are comparable with respect to composition and microstructure, therefore the oxidation behaviour should be similar. Indeed it was seen from the measurements of TGO thickness after isothermal oxidation, that the oxide growth rates can be compared. Furthermore, the static oxidation test performed in parallel to the TCF test also indicated that the two coating systems with low surface roughness were inferior to the two coating systems with larger surface roughness from an integrity point of view.

An explanation based on an increased stress state due to oxidation, and early failure for the two smoothest isothermal oxidation cases, is plausible if the work by Jinnestrand and Sjöström is considered. The smoother TBC system will have a higher opening (tear) stress intensity at the interface than a coarser geometry when a thick TGO is considered during a major portion of the coating fatigue life. Therefore it is possible to state that the difference in damage development for the four cases does not mainly stem from large variations in oxidation behaviour, but rather the stress state introduced by the oxide and geometry.

By comparing Figure 5 and Figure 10 it is obvious that life assessment data are not easily interpreted, due to the fact that the coarser interface at a first glance all give equal results and are inferior to the 10/70 interface. In Figure 5 it is suggested from tests that an increased surface roughness has a strong positive effect on fatigue life. Model prediction results need to be studied more in detail, since the trend is not obvious. A positive effect on fatigue life of increasing the surface roughness is found for a smooth interface. When the surface roughness is increased from the 10/70 level, the predicted fatigue life is shortened and it appears as if the effect of surface roughness on fatigue life is not as significant. However, when comparing only the 20/70 and 30/70-cases, damage seems to initiate earlier if the interface asperity is increased. Growth of existing cracks on the other hand, seems, according to the prediction, to take place more easily in the 20/70 case in comparison to the 30/70 case. This is in agreement with the findings in the experimental work.

The effect of surface roughness on TBC spallation life has been addressed in the literature. Khan and Lu [28] showed that an increased surface roughness would give a better resistance to TBC spallation. Scrivani et.al. [29] performed TCF testing on thick TBC systems and found a similar result. Liu et.al. [30] pointed out that a large surface roughness (13μm) or a very smooth interface (4nm) can both give a better fatigue life than an intermediate surface roughness provided that a good bonding, i.e. an adherent TGO, between BC and TC is allowed to form. In the present work it is shown experimentally that an increased surface roughness has a beneficial effect on fatigue life. Modelling results capture the same trends when surface roughness data for w_a = 6μm and 12μm are compared. However, a life assessment for a smoother interface (w_a = 3μm) shows a different trend, a flat interface is even more beneficial from a fatigue life point of view. It is likely that a smooth interface will be beneficial as long as defects and local irregularities are not present. As soon as an irregularity is present, it will act as a stress concentration for crack formation. Growth of interface cracks can then take place, especially when the TGO is allowed to grow and increase the stress intensity at the interface further. The effect of bond strength at the interface is not considered here.

It seems as if the model can capture the observed damage development behaviour, even though the model calibration is not performed on more than one interface geometry, indicting that the factors influencing the relation K_I / K_{II}, namely λ_0 and m, most likely can be improved further, especially for a very smooth interface, where the shear loading has less influence on crack growth. Also a statistical approach to interface geometry can be useful in describing damage development at the TC/BC interface.

CONCLUSIONS

In the present work a study is performed, where measured and modelled TBC spallation life data are compared. Observations of the failure surface indicates that the fracture surface is located at the TGO / TC interface and that the spallation failure can be considered a black failure, i.e. failure within the thermally grown oxide. An analysis of the formed TC/BC interface oxide shows that the TGO mainly consists of alumina with yttrium-aluminium-oxide stringers at the point of failure. In addition is chromia observed in the TGO at the TGO/TC interface.

Thermal fatigue testing reveals that an increased surface roughness is beneficial from a fatigue life point of view. The same is also observed in life predictions of corresponding geometries. In the present experimental work, the interface roughness was varied between 6-11μm. For a smoother interface the fatigue life model predicts that the fatigue life is increased, this was, however, not experimentally verified.

ACKNOWLEDGEMENTS

This research has partly been funded by the Swedish Energy Agency, Siemens Industrial Turbomachinery AB and Volvo Aero Corporation through the Swedish research program TURBO POWER, the support of which is gratefully acknowledged.

REFERENCES

[1] Eyre, B.L., Progress Mater Sci, 42, (1997), pp. 23-37
[2] Sehra, A.K. and Whitlow, W., Progress Aerospace Sci, 40, (2004), pp. 199-235
[3] Meijer, S.M. et.al., Thermal barrier coating life prediction model development Phase II, Final report, NASA CR 18911, (1991)
[4] Evans, A.G. and Hutchinson, J.W., Int J Solids Structures, 20, (1994), pp. 445-466
[5] Busso, E.P. et .al., Acta Mater, 55, (2007), pp. 1491-1503
[6] Bose, S., High temperature coatings, Butterworth-Heinemann, (2007)
[7] Rösler, J. et.al., Acta Mater, 49, (2001), pp. 3659-3670
[8] Renusch, D. et.al., Mater Corr, 59, (2008), pp. 1-10
[9] Sohn, Y.H. et.al., J Mater Eng Perform, 3, (1994), pp. 55-60
[10] Chen, W.R. et.al., Surf Coat Technol, 202, (2008), pp. 2677-2683
[11] Echsler, H. et.al., Mater Sci Technol, 20, (2004), pp. 307-318
[12] Aktaa, J. et.al., Acta Mater, 53, (2005), pp. 4399-4413
[13] Wang, J.S. and Evans, A.G., Acta Mater, 47, (1999), pp. 699-710
[14] Renusch, D. et.al., Mater High Temp, 21, (2004), pp. 1-10
[15] Brodin, H., Failure of thermal barrier coatings under thermal and mechanical fatigue loading, Linköping studies in science and technology, Dissertation No 898, (2004), Linköping

[16] Jinnestrand, M. Delamination in APS applied thermal barrier coatings: Life modelling, Linköping studies in science and technology, Dissertation No 902, (2004), Linköping
[17] Jinnestrand, M. and Sjöström, S., Surf Coat Technol, 135, (2001), pp. 188-195
[18] Casu, A. et.al., In: Advanced ceramic coatings and interfaces, Edited by: Zhu, D. and Schulz U., 30th International conference on advanced ceramics and composites, January 22-27, 2006, Cocoa Beach, FL, pp. 115-126
[19] Jinnestrand, M. and Brodin, H., Mater Sci Eng A, 379, (2004), pp. 45-57
[20] Casu, A. et.al., Key Eng Mater, 333, (2007), pp. 263-268
[21] Brodin, H., In: 3:rd International surfaceengieering conference, August 2-4, (2004), Orlando, FL
[22] Brodin, H. and Li, X.H., In: proceedings of the international thermal spray conference and exposition, may 10-12, (2004), Osaka, Japan
[23] Brodin, H. et.al., Damage development in two thermal barrier coating systems, To be presented at the 12th International Conference on Fracture, July 12-17, (2009), Ottawa, Canada
[24] Haynes, J.A. et.al., J Thermal Spray Technol, 9, (2000), pp. 38-48
[25] Hutchinson, J.W. and Suo, Z., Adv Appl Mech, 29, (1990), pp. 63-187
[26] Renusch D. et.al., Mater High Temp, 21, (2004), pp. 65-76
[27] Yan, J., Acta Mater, 56, (2008), pp. 4080-4090
[28] Khan, A.N. and Lu, J., surf Coat Technol, 201, (2007), pp. 4653-4658
[29] Scrivani. A. et.al., Mater Sci Eng A, 476, (2008), pp. 1-7
[30] Liu, J. et.al., Surf Coat Technol, 200, (2006), pp. 5869-5876

Author Index

Printed in the United States
By Bookmasters